Elliott H. Lieb
Robert Seiringer
Jan Philip Solovej
Jakob Yngvason

The Mathematics of the Bose Gas and its Condensation

Birkhäuser Verlag
Basel · Boston · Berlin

Authors:

Elliott H. Lieb
Departments of Mathematics and Physics
Princeton University
Jadwin Hall, P.O. Box 708
Princeton, NJ 08544
USA
lieb@math.princeton.edu

Robert Seiringer
Department of Physics
Princeton University
Jadwin Hall, P.O. Box 708
Princeton, NJ 08544
USA
rseiring@math.princeton.edu

Jan Philip Solovej
Department of Mathematics
University of Copenhagen
Universitetsparken 5
2100 Copenhagen
Denmark
solovej@math.ku.dk

Jakob Yngvason
Institut für Theoretische Physik
Universität Wien
Boltzmanngasse 5
1090 Wien
Austria
yngvason@thor.thp.univie.ac.at

2000 Mathematical Subject Classification 82Bxx

A CIP catalogue record for this book is available from the
Library of Congress, Washington D.C., USA

Bibliografische Information Der Deutschen Bibliothek
Die Deutsche Bibliothek verzeichnet diese Publikation in der Deutschen Nationalbibliografie;
detaillierte bibliografische Daten sind im Internet über <http://dnb.ddb.de> abrufbar.

ISBN 3-7643-7336-9 Birkhäuser Verlag, Basel – Boston – Berlin

© 2005 by Elliott H. Lieb, Robert Seiringer, Jan Philip Solovej, Jakob Yngvason
Published by Birkhäuser Verlag, P.O. Box 133, CH-4010 Basel, Switzerland
Part of Springer Science+Business Media
Cover design: Micha Lotrovsky, CH-4106 Therwil, Switzerland
Printed on acid-free paper produced from chlorine-free pulp. TCF ∞

ISBN-10: 3-7643-7336-9
ISBN-13: 978-3-7643-7336-8

9 8 7 6 5 4 3 2 1

www.birkhauser.ch

Contents

Preface

The mathematical study of the Bose gas goes back to the first quarter of the twentieth century, with the invention of quantum mechanics. The name refers to the Indian physicist S.N. Bose who realized in 1924 that the statistics governing photons (essentially invented by Max Planck in 1900) is determined (using modern terminology) by restricting the physical Hilbert space to be the symmetric tensor product of single photon states. Shortly afterwards, Einstein applied this idea to massive particles, such as a gas of atoms, and discovered the phenomenon that we now call Bose-Einstein condensation. At that time this was viewed as a mathematical curiosity with little experimental interest, however.

The peculiar properties of liquid Helium (first liquefied by Kammerlingh Onnes in 1908) were eventually viewed as an experimental realization of Bose-Einstein statistics applied to Helium atoms. The unresolved mathematical problem was that the atoms in liquid Helium are far from the kind of non-interacting particles envisaged in Einstein's theory, and the question that needed to be resolved was whether Bose-Einstein condensation really takes place in a strongly interacting system — or even in a weakly interacting system.

That question is still with us, three quarters of a century later!

The first systematic and semi-rigorous mathematical treatment of the problem was due to Bogoliubov in 1947, but that theory, while intuitively appealing and undoubtedly correct in many aspects, has major gaps and some flaws. The 1950's and 1960's brought a renewed flurry of interest in the question, but while theoretical intuition benefited hugely from this activity the mathematical structure did not significantly improve.

The subject was largely quiescent until the 1990's when experiments on low density (and, therefore, weakly interacting instead of strongly interacting, as in the case of liquid Helium) gases showed for the first time an unambiguous manifestation of Bose-Einstein condensation. This created an explosion of activity in the physics community as can be seen from the web site
`http://bec01.phy.georgiasouthern.edu/bec.html/bibliography.html`,
which contains a bibliography of several thousand papers related to BEC written in the last 10 years.

At more or less the same time some progress was made in obtaining rigorous mathematical proofs of some of the properties proposed in the 50's and 60's. A general proof of Bose-Einstein condensation for interacting gases still eludes us, but we are now in a much stronger position to attack this problem rigorously. These notes, which are an extension of our 2004 Oberwolfach course, summarize some rigorous results that have been obtained by us in the past decade. Most of them are about the ground state energy in various models and dimensions, but we do have a few results about the occurrence (and non-occurrence) of condensation.

This pedagogical summary has several antecedents. It has grown organically as new results emerged. The first one was [LY3], followed by [LSeY3], [Se2], [L7], [LSeY4], [LSSY], and [LSSY2]. Apart from this stream, there was another pedagogical survey going back to the 60's [L3] that dealt with Bogoliubov theory and other things. Some of that material is reproduced in Appendices A and B.

There is, of course, a large body of rigorous work by other people on various aspects of BEC that was not covered in the Oberwolfach course and is not mentioned in these notes. The subject can be approached from many angles and our aim was not to give a complete overview of the subject but to focus on themes where we have been able to make some contributions. The recent Physics Reports article [ZB] on the Bogoliubov model is a good source of references to some other approaches and results. There exist also several reviews, e.g., [DGPS, ISW, Leg, C, Yu] and even monographs [PS, PiSt2] on the fascinating physics of the Bose gas and its condensation.

Acknowledgments

Our thanks go to Letizzia Wastavino for producing the two figures and to Michael Aizenman and Kai Schnee for permission to include the material in Chapters 11 and 9, respectively. We are also grateful to Daniel Ueltschi, Bruno Nachtergaele and Valentin Zagrebnov for very useful comments on various parts of these notes. Finally, we thank the Oberwolfach Institute for the opportunity to give this course and we thank the participants for pointing out numerous typos in a previous version of the manuscript. The work was supported in part by US NSF grants PHY 0139984-A01 (EHL), PHY 0353181 (RS) and DMS-0111298 (JPS); by an A.P. Sloan Fellowship (RS); by EU grant HPRN-CT-2002-00277 (JPS and JY); by the Alexander von Humboldt Foundation (EHL); by FWF grant P17176-N02 (JY); by MaPhySto – A Network in Mathematical Physics and Stochastics funded by The Danish National Research Foundation (JPS), and by grants from the Danish research council (JPS).

Chapter 1

Introduction

1.1 The Ideal Bose Gas

Schrödinger's equation of 1926 defined a new mechanics whose Hamiltonian is based on classical mechanics. The "ideal" gas of particles consists of the following ingredients: A collection of $N \gg 1$ non-interacting particles in a large box $\Lambda \subset \mathbb{R}^3$ and volume $V = L^3$. We are interested in the "thermodynamic limit", which means that we will take $N \to \infty$ and $L \to \infty$ in such a way that the density $\rho = N/V$ is held fixed.

The fact that the particles are non-interacting means that the classical energy, or Hamiltonian H, is entirely kinetic energy and this, in turn, is

$$H = \frac{1}{2m} \sum_{i=1}^{N} \mathbf{p}_i^2 \tag{1.1}$$

where m is the particle mass and $\mathbf{p}_i$ is the momentum of particle i. The lowest (or *ground state*) energy of this classical system is, of course, 0. The *thermodynamic properties* are determined from the *partition function*

$$Z_N = \frac{1}{h^{3N} N!} \int_{\Lambda^N} \prod_{i=1}^{N} d\mathbf{x}_i \int_{\mathbb{R}^{3N}} \prod_{i=1}^{N} d\mathbf{p}_i \, e^{-\beta H} = \frac{V^N}{h^{3N} N!} \left(\frac{2m}{\beta} \right)^{3N/2} \tag{1.2}$$

(with $\beta = 1/k_B T$, k_B is Boltzmann's constant and T is the temperature, and with h an arbitrary constant with the dimension of momentum times length) in terms of which the *free energy* is given as

$$F = -\frac{1}{\beta} \ln Z_N. \tag{1.3}$$

The pressure p is

$$p = -\partial F / \partial V = \rho k_B T \tag{1.4}$$

where Stirling's approximation, $\ln N! \approx N \ln N$, has been used. The average energy is

$$E = -\frac{\partial}{\partial \beta} \ln Z = \frac{3}{2} N k_B T. \tag{1.5}$$

In quantum mechanics, the Hamiltonian is an operator obtained by replacing each $\mathbf{p}_j$ by $-i\hbar \nabla_j$, acting on the Hilbert space $L^2(\Lambda)$, with appropriate boundary conditions. The eigenvalues of $\mathbf{p}^2 = -\hbar^2 \Delta$ (with $\Delta =$ Laplacian $=\nabla^2$), for a box with periodic boundary conditions, are $(2\pi\hbar)^2 \mathbf{n}^2 / L^2$, where $\mathbf{n}$ is a vector with integer components. If the statistics of the particles is disregarded, the partition function, which in the quantum case is given by

$$Z_N = \frac{1}{N!} \operatorname{tr} e^{-\beta H}, \tag{1.6}$$

factorizes as $Z_N = Z_1^N / N!$. This equals the classical expression in the thermodynamic limit, if one takes $h = 2\pi\hbar$.

Taking the statistics of the particles into account, we have to restrict the trace in (1.6) to the symmetric or anti-symmetric subspace of the total Hilbert space $\bigotimes^N L^2(\Lambda)$, depending on whether we intend to describe bosons or fermions. This makes the prefactor $1/N!$ superfluous, but one has to face the problem that Z_N is no longer determined by Z_1. For this reason, it is more convenient to pass to the grand-canonical partition function (or generating function)

$$\Xi = \sum_{N \geq 0} Z_N z^N, \tag{1.7}$$

where $z = e^{\beta\mu}$ is the fugacity for chemical potential μ. The chemical potential is then determined by the average particle number,

$$\langle N \rangle = z \frac{\partial}{\partial z} \ln \Xi. \tag{1.8}$$

The grand-canonical partition function Ξ can be calculated because it factorizes into the contributions from the single particle energy levels. For bosons the result is

$$\Xi = \prod_{i \geq 0} \frac{1}{1 - \exp\left(-\beta(\varepsilon_i - \mu)\right)}, \tag{1.9}$$

where $\varepsilon_0 \leq \varepsilon_1 \leq \ldots$ denote the single-particle energy levels (given, in this case, by $(2\pi\hbar)^2 \mathbf{n}^2 / (2mL^2)$, with $\mathbf{n} \in \mathbb{Z}^3$). Note that in this "free particle" bosonic case it is necessary that $\mu < \varepsilon_0$. In the thermodynamic limit $\varepsilon_0 \to 0$.

For fixed $\mu < 0$, the average particle number, is given by

$$\langle N \rangle = \sum_{i \geq 0} \frac{1}{\exp\left(\beta(\varepsilon_i - \mu)\right) - 1}. \tag{1.10}$$

In the thermodynamic limit, the sum becomes an integral (more precisely, $L^{-3} \sum_{\mathbf{p}} \to (2\pi\hbar)^{-3} \int_{\mathbb{R}^3}$), and we have

$$\lim_{L \to \infty} \frac{\langle N \rangle}{L^3} \equiv \rho = h^{-3} \int d\mathbf{p} \, \frac{1}{\exp\left(\beta(\mathbf{p}^2/(2m) - \mu)\right) - 1}. \tag{1.11}$$

This is a monotonously increasing function of μ, which is bounded as $\mu \to 0$, however, by the *critical density*

$$\rho_c(\beta) = g_{3/2}(1)(2\pi\hbar^2\beta/m)^{-3/2}. \tag{1.12}$$

Here $g_{3/2}(1) = \sum_{\ell=1}^{\infty} \ell^{-3/2} \approx 2.612$. That is, the density seems to be bounded by this value. This is absurd, of course. This phenomenon was discovered by Einstein [E], and the resolution is that the particles exceeding the critical number all go into the lowest energy state. In mathematical terms, this means that we have to let $\mu \to 0$ *simultaneously* with $L \to \infty$ to fix the density at some number $> \rho_c$. In this case, we have to be more careful in replacing the sum in (1.10) by an integral. It turns out to be sufficient to separate the contribution from the lowest energy level, and approximate the contribution from the remaining terms by an integral. The result is that, for $\rho > \rho_c$,

$$\rho = \rho_c(\beta) + \rho_0, \tag{1.13}$$

where

$$\rho_0 = \lim_{L \to \infty} \frac{1}{V} \frac{1}{\exp\left(\beta(\varepsilon_0 - \mu)\right) - 1} \tag{1.14}$$

is the density of the "condensate". The dependence of μ on L is determined by (1.10), writing $\langle N \rangle = L^3\rho$ with fixed ρ.

The phenomenon that a single particle level has a macroscopic occupation, i.e., a non-zero density in the thermodynamic limit, is called *Bose-Einstein condensation* (BEC). Note that in the model considered there is no condensation into the excited energy levels, and one always has

$$\lim_{L \to \infty} \frac{1}{V} \frac{1}{\exp\left(\beta(\varepsilon_i - \mu)\right) - 1} = 0 \tag{1.15}$$

for $i \geq 1$, since $\varepsilon_i - \mu \geq \varepsilon_i - \varepsilon_0 = \text{const.} \, L^{-2}$.

Note that in the case of zero temperature, i.e., the ground state, *all* the particles are in the condensate, i.e., $\rho = \rho_0$. In a sector of fixed particle number, the ground state wave function is simply a product of single particle wave-functions in the lowest energy state.

1.2   The Concept of Bose-Einstein Condensation

So far we have merely reproduced the standard textbook discussion of BEC for non-interacting particles. The situation changes drastically if one considers inter-acting systems, however. For particles interacting via a pair potential $v(|\mathbf{x}_i - \mathbf{x}_j|)$, the Hamiltonian takes the form

$$H_N = -\frac{\hbar^2}{2m}\sum_{i=1}^{N}\Delta_i + \sum_{1 \leq i < j \leq N} v(|\mathbf{x}_i - \mathbf{x}_j|). \tag{1.16}$$

(We could also include three- and higher body potentials, but we exclude them for simplicity. These do exist among real atoms, but for understanding the basic physics it is presumably sufficient to consider only pair potentials.)

Even at zero temperature, it is not entirely obvious what is meant by a macroscopic occupation of a one-particle state, because the eigenfunctions of H_N are *not* products of single particle states.

The concept of a macroscopic occupation of a single one-particle state ac-quires a precise meaning through the *one-particle density matrix*. Given the nor-malized ground state wave function of H_N (or any other wave function, for that matter), Ψ_0, this is the operator on $L^2(\mathbb{R}^3)$ given by the kernel

$$\gamma(\mathbf{x}, \mathbf{x}') = N \int \Psi_0(\mathbf{x}, \mathbf{X})\Psi_0(\mathbf{x}', \mathbf{X})d\mathbf{X} , \tag{1.17}$$

where we introduced the short hand notation

$$\mathbf{X} = (\mathbf{x}_2, \ldots, \mathbf{x}_N) \quad \text{and} \quad d\mathbf{X} = \prod_{j=2}^{N} d\mathbf{x}_j. \tag{1.18}$$

Then $\int \gamma(\mathbf{x}, \mathbf{x})d\mathbf{x} = \mathrm{tr}[\gamma] = N$. *BEC in the ground state means, by definition, that this operator has an eigenvalue of order N in the thermodynamic limit.* This formulation was first stated in [PO] by Penrose and Onsager. For the ground state Ψ_0 of H_N, the kernel γ is positive and, hopefully, translation invariant in the thermodynamic limit, and hence the eigenfunction belonging to the largest eigenvalue must be the constant function $L^{-3/2}$. Therefore, another way to say that there is BEC in the ground state is that

$$\frac{1}{V} \iint \gamma(\mathbf{x}, \mathbf{y}) \, d\mathbf{x} \, d\mathbf{y} = O(N) \tag{1.19}$$

as $N \to \infty$, $L \to \infty$ with N/L^3 fixed; more precisely Eq. (1.19) requires that there is a $c > 0$ such that the left side is $> cN$ for all large N.

This concept of BEC as a large eigenvalue of the one-particle reduced density matrix immediately generalizes to thermal states, both in the canonical and grand-canonical ensembles (or, more generally, to states defined by arbitrary density matrices). For interacting systems, however, the ground state already poses a challenging problem which is still largely unsolved. In fact, BEC has, so far, never been proved for many-body Hamiltonians with genuine interactions — *except for one special case:* hard core bosons on a lattice at half-filling (i.e., N = half the number of lattice sites). The proof was given in [DLS] and [KLS]. This, and a generalization to a lattice gas in a periodic external potential, is described in Chapter 11.

There are physical situations where it is natural to consider generalizations of the concept of BEC just described. In particular, for trapped Bose gases, as considered in Chapters 6–9, the system is inhomogeneous and the thermodynamic limit at fixed density has to be replaced by an appropriate scaling of the potentials involved.[1] What remains as the criterion of BEC is the occurrence of a large eigenvalue (i.e., of the order of the particle number) of the one-particle density matrix in the limit considered.

Finally, we comment on the relation between BEC and *spontaneous breaking of gauge symmetry.* Gauge symmetry in the present context is defined by the one-parameter group of unitary transformations in Fock space generated by the particle number operator. The Hamiltonian (1.16) preserves particle number and this implies gauge invariance of the grand-canonical equilibrium state on the Fock space, i.e., the direct sum of all the N-particle Hilbert spaces. This symmetry can be explicitly broken by adding a term $\sqrt{V}\lambda(a_0 + a_0^*)$ to the Hamiltonian, where a_0^* and a_0 are the creation and annihilation operators of the lowest energy mode (i.e., in a box, the constant wave function in the one-particle space). Formally, this

---

[1] Also, the way to take a thermodynamic limit is not completely unambiguous. In the standard van Hove limit [Ru] the volume grows essentially without changing the shape of the domain Λ, e.g., the growing cubes considered in Section 1.1. If instead the limit is taken in such a way that the size of Λ grows at different speed in different directions, one may obtain different results (see, e.g., [BL, BLP]).

is analogous to adding an external magnetic field to the Hamiltonian of a magnet. In the grand-canonical state defined by the so modified Hamiltonian the operator a_0 has a non-zero expectation value (which goes to zero as $\lambda \to 0$ for any *fixed* volume V). Gauge symmetry breaking means that this expectation value, divided by $\sqrt{V}$, remains non-zero even as $\lambda \to 0$, *after* the thermodynamic limit has been taken. (For the magnet, this corresponds to spontaneous magnetization.)

In Appendix D, we show that under quite general assumptions, BEC goes hand in hand with spontaneous gauge symmetry breaking. Breaking of a continuous symmetry is notoriously difficult to prove, and in one and two dimensions it is excluded, at least at positive temperature, by the Hohenberg-Mermin-Wagner Theorem [Ho, MW]. This partly explains why a rigorous proof of BEC for interacting systems is still lacking in general.

1.3   Overview and Outline

A central theme of these notes is the evaluation of the ground state energy in various situations. Already in 1927 W. Lenz [Len] gave a heuristic argument indicating that the ground state energy of a dilute hard-core gas was proportional to the scattering length, a, of the potential in three dimensions.[2] The 3D formula for the energy per particle at low density is $e_0(\rho) = (\hbar^2/2m)4\pi\rho a$, to leading order in the density $\rho = N/V$. Bogoliubov's 1947 paper [Bo] showed how a perturbative version of this formula could be derived in a more systematic and general way, and there was further work on this in the 1950's and 60's [HY, L2], but it was not until 1998 [LY1] that it was proved rigorously. The issue here was to derive a lower bound for the energy. In 1957 Dyson [D1] had established an asymptotically correct upper bound for the energy and also a lower bound, but the latter was 14 times too small.

In two dimensions a different formula, postulated as late as 1971 by Schick [S], holds and was rigorously proved to be correct in [LY2]. With the aid of the methodology developed to prove the lower bound for the homogeneous gas, several other problems could successfully be addressed. One is the proof that the Gross-Pitaevskii equation correctly describes the ground state in the 'traps' actually

---

[2]Lenz calculated the ground state energy of a single particle in the presence of $N-1$ randomly placed particles in a $L \times L \times L$ box. This would give the correct energy, $4\pi a/V$, *provided* the fixed particles are uniformly distributed. But in a random arrangement there are bound to be holes of arbitrary large size (as $N \to \infty$) and the moving particle can always fit in a hole with arbitrary small energy. It is the inevitability of such holes that makes it impossible to convert Lenz's argument into a rigorous one.

used in the experiments [LSeY1, LSeY2]. For such systems it is also possible to prove complete Bose condensation and superfluidity [LSe, LSeY5]. On the frontier of experimental developments is the possibility that a dilute gas in an elongated trap will behave like a one-dimensional system and this topic was addressed in a mathematical way in [LSeY6].

Another topic is a proof that Foldy's 1961 theory [F] of a *high density* Bose gas of charged particles, which is based on Bogoliubov's theory, correctly describes the ground state energy of this gas; using this it is also possible to prove the $N^{7/5}$ formula for the ground state energy of the two-component charged Bose gas proposed by Dyson in 1967 [D2]. All of this is quite recent work [LSo, LSo2, So] and it is hoped that the mathematical methodology might be useful, ultimately, to solve more complex problems connected with these interesting systems.

One of the most remarkable recent developments in the study of ultracold Bose gases is the observation of a reversible transition from a Bose-Einstein condensate to a state composed of localized atoms as the strength of a periodic, optical trapping potential is varied [JBCGZ, G2, G3]. Together with M. Aizenman we have rigorously analyzed a model of this phenomenon [ALSSY]. The gas is a hard core lattice gas and the optical lattice is modeled by a periodic potential of strength λ. For small λ and temperature BEC is proved to occur, while at large λ BEC disappears, even in the ground state, which is a Mott insulator state with a characteristic gap. The inter-particle interaction is essential for this effect.

Let us briefly describe the structure of these notes. The discussion centers around six main topics:

1. The dilute, homogeneous Bose gas with repulsive interaction (2D and 3D).
2. Repulsive bosons in a trap (as used in recent experiments) and the "Gross-Pitaevskii" equation.
3. BEC and superfluidity for dilute trapped gases.
4. Low-dimensional behavior of three-dimensional gases in elongated or disc-shaped traps.
5. Foldy's "jellium" model of charged particles in a neutralizing background and the extension to the two-component gas.
6. The model of an optical lattice that shows BEC at weak coupling and no BEC at strong coupling.

The discussion below of topic 1 is based on [LY1] and [LY2], and of topic 2 on [LSeY1] and [LSeY2]. See also [LY3, LSeY3, Se2, LSeY4].

The original references for topic 3 are [LSe] and [LSeY5], but for transparency we also include here a chapter on the special case when the trap is a rectangular box. This case already contains the salient points, but avoids several complications due the inhomogeneity of the gas in a general trap. An essential technical tool for topic 3 is a generalized Poincaré inequality, which is discussed in a separate chapter.

Topic 4 is a summary of the work in [LSeY6] and [SY].

The discussion of topic 5 is based on [LSo], [LSo2] and [So]. Topic 6 is based on [ALSSY] and [ALSSY2].

Topic 1 (3 dimensions) was the starting point and contains essential ideas. It is explained here in some detail and is taken, with minor modifications (and corrections), from [LY3]. In terms of technical complexity, however, the fifth topic is the most involved and can not be treated here in full detail.

The interaction potential between pairs of particles in the Jellium model and the two-component gas in topic 5 is the repulsive, *long-range* Coulomb potential, while in topics 1–4 it is assumed to be repulsive and *short range*. Topic 6 concerns a hard-core lattice gas. For alkali atoms in the recent experiments on Bose-Einstein condensation the interaction potential has a repulsive hard core, but also a quite deep attractive contribution of van der Waals type and there are many two body bound states [PS]. The Bose condensate seen in the experiments is thus not the true ground state (which would be a solid) but a metastable state. Nevertheless, it is usual to model this metastable state as the ground state of a system with a repulsive two body potential having the same scattering length as the true potential, and this is what we shall do. In these notes all interaction potentials will be positive, except in Appendices A, C and D.

There are four appendices to these notes:

A. Elements of Bogoliubov Theory

B. An Exactly Soluble Model

C. Definition and Properties of the Scattering Length

D. *c*-Number Substitutions and Gauge Symmetry Breaking

Appendices A and B are reproductions, with only minor modifications, of two sections of the survey article [L3]. This material is still relevant after 40 years and parts of it were discussed in the Oberwolfach course. Appendix C is taken from the paper [LY2]. Appendix D is a slightly extended version of [LSeY8].

Chapter 2

The Dilute Bose Gas in 3D

We consider the Hamiltonian for N bosons of mass m enclosed in a cubic box Λ of side length L and interacting by a spherically symmetric pair potential $v(|\mathbf{x}_i - \mathbf{x}_j|)$:

$$H_N = -\mu \sum_{i=1}^{N} \Delta_i + \sum_{1 \leq i < j \leq N} v(|\mathbf{x}_i - \mathbf{x}_j|). \tag{2.1}$$

Here $\mathbf{x}_i \in \mathbb{R}^3$, $i = 1, \ldots, N$ are the positions of the particles, Δ_i the Laplacian with respect to $\mathbf{x}_i$, and we have denoted $\hbar^2/2m$ by μ for short. (By choosing suitable units μ could, of course, be eliminated, but we want to keep track of the dependence of the energy on Planck's constant and the mass.) The interaction potential will be assumed to be *nonnegative* and to decrease faster than $1/r^3$ at infinity. Note that for potentials that tend to zero at infinity 'repulsive' and 'non-negative' are synonymous — in the quantum mechanical literature at least. In classical mechanics, in contrast, a potential that is positive but not monotonically decreasing is not called repulsive.

The Hamiltonian (2.1) operates on *symmetric* wave functions in the Hilbert space $L^2(\Lambda^N, d\mathbf{x}_1 \cdots d\mathbf{x}_N)$ as is appropriate for bosons. Let us note an important fact here that will be useful later. To say that we are in the symmetric tensor product of $L^2(\mathbb{R}^3)$ is equivalent, in plain language, to the statement that we consider only wave functions $\Psi(\mathbf{x}_1, \ldots, \mathbf{x}_N)$ that are invariant under permutation of the N coordinates. On the other hand, we could ask for the bottom of the spectrum of H_N without imposing this symmetry restriction, i.e., on the whole tensor product. The fact is that the two are the same. The absolute ground state energy is the boson ground state energy. Moreover, the ground state is unique, i.e., there is only one and it is symmetric. This fact will be useful later because it means that we can get an upper bound to the ground state energy by using any handy function.

The proof of this assertion, very briefly, goes as follows: The absolute ground state energy is the infimum of $\langle \Psi | H_N | \Psi \rangle$ over all normalized Ψ. If Ψ is a candidate for a minimizer, we can consider $\Phi(\mathbf{X}) = |\Psi(\mathbf{X})|$, which has the same norm, the same potential energy, and a kinetic energy that can be lower but not higher [LLo]. This Φ will not satisfy the Schrödinger equation unless $\Phi = \Psi$, up to an overall phase. Thus, the absolute ground state is unique and it has to be either symmetric or antisymmetric since these are the only one-dimensional representations of the permutation group. Since it has only one sign it is symmetric.

We are interested in the ground state energy $E_0(N, L)$ of (2.1) in the *thermodynamic limit* when N and L tend to infinity with the density $\rho = N/L^3$ fixed. The energy per particle in this limit is

$$e_0(\rho) = \lim_{L \to \infty} E_0(\rho L^3, L)/(\rho L^3). \tag{2.2}$$

Our results about $e_0(\rho)$ are based on estimates on $E_0(N, L)$ for finite N and L, which are important, e.g., for the considerations of inhomogeneous systems in Chapters 6–9. To define $E_0(N, L)$ precisely one must specify the boundary conditions. These should not matter for the thermodynamic limit. To be on the safe side we use Neumann boundary conditions for the lower bound, and Dirichlet boundary conditions for the upper bound since these lead, respectively, to the lowest and the highest energies.

For experiments with dilute gases the *low density asymptotics* of $e_0(\rho)$ is of importance. Low density means here that the mean interparticle distance, $\rho^{-1/3}$ is much larger than the *scattering length* a of the potential, which is defined as follows. (See Appendix C for details.) The zero energy scattering Schrödinger equation

$$-2\mu \Delta \psi + v(r)\psi = 0 \tag{2.3}$$

has a solution of the form, asymptotically as $|\mathbf{x}| = r \to \infty$ (or for all $r > R_0$ if $v(r) = 0$ for $r > R_0$),

$$\psi_0(\mathbf{x}) = 1 - a/|\mathbf{x}|. \tag{2.4}$$

(The factor 2 in (2.3) comes from the reduced mass of the two particle problem.) Writing $\psi_0(\mathbf{x}) = u_0(|\mathbf{x}|)/|\mathbf{x}|$ this is the same as

$$a = \lim_{r \to \infty} r - \frac{u_0(r)}{u_0'(r)}, \tag{2.5}$$

where u_0 solves the zero energy (radial) scattering equation,

$$-2\mu u_0''(r) + v(r)u_0(r) = 0 \tag{2.6}$$

with $u_0(0) = 0$.

An important special case is the hard core potential $v(r) = \infty$ if $r < a$ and $v(r) = 0$ otherwise. Then the scattering length and the radius a are the same.

Our main result is a rigorous proof of the formula

$$e_0(\rho) \approx 4\pi\mu\rho a \tag{2.7}$$

for $\rho a^3 \ll 1$, more precisely of

Theorem 2.1 (Low density limit of the ground state energy).

$$\lim_{\rho a^3 \to 0} \frac{e_0(\rho)}{4\pi\mu\rho a} = 1. \tag{2.8}$$

This formula is independent of the boundary conditions used for the definition of $e_0(\rho)$. It holds for every positive radially symmetric pair potential such that $\int_R^\infty v(r)r^2 dr < \infty$ for some R, which guarantees a finite scattering length, cf. Appendix C.

The genesis of an understanding of $e_0(\rho)$ was the pioneering work [Bo, Bo3] of Bogoliubov, and in the 50's and early 60's several derivations of (2.8) were presented [HY], [L2], even including higher order terms:

$$\frac{e_0(\rho)}{4\pi\mu\rho a} = 1 + \frac{128}{15\sqrt{\pi}}(\rho a^3)^{1/2} + 8\left(\frac{4\pi}{3} - \sqrt{3}\right)(\rho a^3)\log(\rho a^3) + O(\rho a^3). \tag{2.9}$$

These early developments are reviewed in [L3]. They all rely on some special assumptions about the ground state that have never been proved, or on the selection of special terms from a perturbation series which likely diverges. The only rigorous estimates of this period were established by Dyson, who derived the following bounds in 1957 for a gas of hard spheres [D1]:

$$\frac{1}{10\sqrt{2}} \le \frac{e_0(\rho)}{4\pi\mu\rho a} \le \frac{1 + 2Y^{1/3}}{(1 - Y^{1/3})^2} \tag{2.10}$$

with $Y = 4\pi\rho a^3/3$. While the upper bound has the asymptotically correct form, the lower bound is off the mark by a factor of about $1/14$. But for about 40 years this was the best lower bound available!

Under the assumption that (2.8) is a correct asymptotic formula for the energy, we see at once that understanding it physically, much less proving it, is not a simple matter. Initially, the problem presents us with two lengths, $a \ll \rho^{-1/3}$ at low density. However, (2.8) presents us with another length generated by the solution to the problem. This length is the de Broglie wavelength, or 'uncertainty principle' length (sometimes called 'healing length')

$$\ell_c \sim (\rho a)^{-1/2}. \tag{2.11}$$

The reason for saying that ℓ_c is the de Broglie wavelength is that in the hard core case all the energy is kinetic (the hard core just imposes a $\psi = 0$ boundary condition whenever the distance between two particles is less than a). By the uncertainty principle, the kinetic energy is proportional to an inverse length squared, namely ℓ_c. We then have the relation (since ρa^3 is small)

$$a \ll \rho^{-1/3} \ll \ell_c \tag{2.12}$$

which implies, physically, that it is impossible to localize the particles relative to each other (even though ρ is small). Bosons in their ground state are therefore 'smeared out' over distances large compared to the mean particle distance and their individuality is entirely lost. They cannot be localized with respect to each other without changing the kinetic energy enormously.

Fermions, on the other hand, prefer to sit in 'private rooms', i.e., ℓ_c is never bigger than $\rho^{-1/3}$ times a fixed factor. In this respect the quantum nature of bosons is much more pronounced than for fermions.

Since (2.8) is a basic result about the Bose gas it is clearly important to derive it rigorously and in reasonable generality, in particular for more general cases than hard spheres. The question immediately arises for which interaction potentials one may expect it to be true. A notable fact is that it *is not true for all* v with $a > 0$, since there are two body potentials with positive scattering length that allow many body bound states. (There are even such potentials without two body bound states but with three body bound states [Ba].) For such potentials (2.8) is clearly false. Our proof, presented in the sequel, works for nonnegative v, but we conjecture that (2.8) holds if $a > 0$ and v has no N-body bound states for any N. The lower bound is, of course, the hardest part, but the upper bound is not altogether trivial either.

Before we start with the estimates a simple computation and some heuristics may be helpful to make (2.8) plausible and motivate the formal proofs.

With ψ_0 the zero energy scattering solution, partial integration, using (2.3) and (2.4), gives, for $R \geq R_0$,

$$\int_{|\mathbf{x}| \leq R} \{2\mu |\nabla \psi_0|^2 + v|\psi_0|^2\} d\mathbf{x} = 8\pi \mu a \left(1 - \frac{a}{R}\right) \rightarrow 8\pi \mu a \quad \text{for } R \rightarrow \infty. \tag{2.13}$$

Moreover, for positive interaction potentials the scattering solution minimizes the quadratic form in (2.13) for each $R \geq R_0$ with the boundary condition $\psi_0(|\mathbf{x}| = R) = (1-a/R)$. Hence the energy $E_0(2, L)$ of two particles in a large box, i.e., $L \gg a$, is approximately $8\pi\mu a/L^3$. If the gas is sufficiently dilute it is not unreasonable to expect that the energy is essentially a sum of all such two particle contributions. Since there are $N(N-1)/2$ pairs, we are thus lead to $E_0(N, L) \approx 4\pi\mu a N(N-1)/L^3$, which gives (2.8) in the thermodynamic limit.

This simple heuristics is far from a rigorous proof, however, especially for the lower bound. In fact, it is rather remarkable that the same asymptotic formula holds both for 'soft' interaction potentials, where perturbation theory can be expected to be a good approximation, and potentials like hard spheres where this is not so. In the former case the ground state is approximately the constant function and the energy is *mostly potential:* According to perturbation theory $E_0(N, L) \approx N(N-1)/(2L^3) \int v(|\mathbf{x}|) d\mathbf{x}$. In particular it is *independent of μ*, i.e. of Planck's constant and mass. Since, however, $\int v(|\mathbf{x}|) d\mathbf{x}$ is the first Born approximation to $8\pi\mu a$ (note that a depends on μ!), this is not in conflict with (2.8). For 'hard' potentials on the other hand, the ground state is *highly correlated*, i.e., it is far from being a product of single particle states. The energy is here *mostly kinetic*, because the wave function is very small where the potential is large. These two quite different regimes, the potential energy dominated one and the kinetic energy dominated one, cannot be distinguished by the low density asymptotics of the energy. Whether they behave differently with respect to other phenomena, e.g., Bose-Einstein condensation, is not known at present.

Bogoliubov's analysis [Bo, Bo3] presupposes the existence of Bose-Einstein condensation. Nevertheless, it is correct (for the energy) for the one-dimensional delta-function Bose gas [LL], despite the fact that there is (presumably) no condensation in that case [PiSt]. It turns out that BE condensation is not really needed in order to understand the energy. As we shall see, 'global' condensation can be replaced by a 'local' condensation on boxes whose size is independent of L. It is this crucial understanding that enables us to prove Theorem 2.1 without having to decide about BE condensation.

An important idea of Dyson was to transform the hard sphere potential into a soft potential at the cost of sacrificing the kinetic energy, i.e., effectively to move from one regime to the other. We shall make use of this idea in our proof of the lower bound below. But first we discuss the simpler upper bound, which relies on other ideas from Dyson's beautiful paper [D1].

2.1 Upper Bound

The following generalization of Dyson's upper bound holds [LSeY1], [Se1]:

Theorem 2.2 (Upper bound). *Let $\rho_1 = (N-1)/L^3$ and $b = (4\pi\rho_1/3)^{-1/3}$. For non-negative potentials v and $b > a$ the ground state energy of (2.1) with periodic boundary conditions satisfies*

$$E_0(N, L)/N \leq 4\pi\mu\rho_1 a \left(1 + \text{const.}\frac{a}{b}\right). \qquad (2.14)$$

Thus in the thermodynamic limit (and for all boundary conditions)

$$\frac{e_0(\rho)}{4\pi\mu\rho a} \leq 1 + \text{const.} Y^{1/3} \qquad (2.15)$$

provided $Y = 4\pi\rho a^3/3 < 1$.

Proof. We first remark that the expectation value of (2.1) with any trial wave function gives an upper bound to the bosonic ground state energy, even if the trial function is not symmetric under permutations of the variables. (See the discussion at the beginning of this chapter.) The reason is that an absolute ground state of the elliptic differential operator (2.1) (i.e., a ground state without symmetry requirement) is a nonnegative function which can be symmetrized without changing the energy because (2.1) is symmetric under permutations. In other words, the absolute ground state energy is the same as the bosonic ground state energy.

Following [D1] we choose a trial function of the following form

$$\Psi(\mathbf{x}_1, \ldots, \mathbf{x}_N) = F_1(\mathbf{x}_1) \cdot F_2(\mathbf{x}_1, \mathbf{x}_2) \cdots F_N(\mathbf{x}_1, \ldots, \mathbf{x}_N). \qquad (2.16)$$

More specifically, $F_1 \equiv 1$ and F_i depends only on the distance of $\mathbf{x}_i$ to its nearest neighbor among the points $\mathbf{x}_1, \ldots, \mathbf{x}_{i-1}$ (taking the periodic boundary into account):

$$F_i(\mathbf{x}_1, \ldots, \mathbf{x}_i) = f(t_i), \quad t_i = \min\left(|\mathbf{x}_i - \mathbf{x}_j|, j = 1, \ldots, i-1\right), \qquad (2.17)$$

with a function f satisfying

$$0 \leq f \leq 1, \quad f' \geq 0. \qquad (2.18)$$

The intuition behind the ansatz (2.16) is that the particles are inserted into the system one at the time, taking into account the particles previously inserted. While such a wave function cannot reproduce all correlations present in the true ground state, it turns out to capture the leading term in the energy for dilute gases. The form (2.16) is computationally easier to handle than an ansatz of the type $\prod_{i<j} f(|\mathbf{x}_i - \mathbf{x}_j|)$, which might appear more natural in view of the heuristic remarks after Eq. (2.13).

The function f is chosen to be

$$
f(r) = \begin{cases} f_0(r)/f_0(b) & \text{for } 0 \le r \le b, \\ 1 & \text{for } r > b, \end{cases} \tag{2.19}
$$

with $f_0(r) = u_0(r)/r$ the zero energy scattering solution defined by (2.6).

We start by computing the kinetic energy of our trial state defined above. Define $\varepsilon_{ik}(\mathbf{x}_1, \ldots, \mathbf{x}_N)$ by

$$
\varepsilon_{ik} = \begin{cases} 1 & \text{for } i = k \\ -1 & \text{for } t_i = |\mathbf{x}_i - \mathbf{x}_k| \\ 0 & \text{otherwise.} \end{cases} \tag{2.20}
$$

Let $\mathbf{n}_i$ be the unit vector in the direction of $\mathbf{x}_i - \mathbf{x}_{j(i)}$, where $\mathbf{x}_{j(i)}$ denotes the nearest neighbor of $\mathbf{x}_i$ among the points $(\mathbf{x}_1, \ldots, \mathbf{x}_{i-1})$. (Note that $j(i)$ really depends on all the points $\mathbf{x}_1, \ldots, \mathbf{x}_i$ and not just on the index i. Except for a set of zero measure, $j(i)$ is unique.) Then

$$
F^{-1} \nabla_k F = \sum_i F_i^{-1} \varepsilon_{ik} \mathbf{n}_i f'(t_i) , \tag{2.21}
$$

and after summation over k we obtain

$$
F^{-2} \sum_k |\nabla_k F|^2 = \sum_{i,j,k} \varepsilon_{ik} \varepsilon_{jk} (\mathbf{n}_i \cdot \mathbf{n}_j) F_i^{-1} F_j^{-1} f'(t_i) f'(t_j)
$$

$$
\le 2 \sum_i F_i^{-2} f'(t_i)^2 + 2 \sum_{k \le i < j} |\varepsilon_{ik} \varepsilon_{jk}| F_i^{-1} F_j^{-1} f'(t_i) f'(t_j) . \tag{2.22}
$$

The expectation value of the Hamiltonian can thus be bounded as follows:

$$
\frac{\langle \Psi | H_N | \Psi \rangle}{\langle \Psi | \Psi \rangle} \le 2\mu \sum_{i=1}^N \frac{\int |\Psi|^2 F_i^{-2} f'(t_i)^2}{\int |\Psi|^2} + \sum_{j<i} \frac{\int |\Psi|^2 v(|\mathbf{x}_i - \mathbf{x}_j|)}{\int |\Psi|^2}
$$

$$
+ 2\mu \sum_{k \le i < j} \frac{\int |\Psi|^2 |\varepsilon_{ik} \varepsilon_{jk}| F_i^{-1} F_j^{-1} f'(t_i) f'(t_j)}{\int |\Psi|^2} . \tag{2.23}
$$

For $i < p$, let $F_{p,i}$ be the value that F_p would take if the point $\mathbf{x}_i$ were omitted from consideration as a possible nearest neighbor. Note that $F_{p,i}$ is independent of $\mathbf{x}_i$. Analogously we define $F_{p,ij}$ by omitting $\mathbf{x}_i$ and $\mathbf{x}_j$. The functions F_i occur both in the numerator and the denominator so we need estimates from below and above. Since f is monotone increasing,

$$F_p = \min\{F_{p,ij}, f(|\mathbf{x}_p - \mathbf{x}_j|), f(|\mathbf{x}_p - \mathbf{x}_i|)\} \,, \tag{2.24}$$

and we have, using $0 \le f \le 1$,

$$F_{p,ij}^2 f(|\mathbf{x}_p - \mathbf{x}_i|)^2 f(|\mathbf{x}_p - \mathbf{x}_j|)^2 \le F_p^2 \le F_{p,ij}^2 \,. \tag{2.25}$$

Hence, for $j < i$, we have the upper bound

$$F_{j+1}^2 \cdots F_{i-1}^2 F_{i+1}^2 \cdots F_N^2 \le F_{j+1,j}^2 \cdots F_{i-1,j}^2 F_{i+1,ij}^2 \cdots F_{N,ij}^2 \,, \tag{2.26}$$

and the lower bound

$$F_j^2 \cdots F_N^2 \ge F_{j+1,j}^2 \cdots F_{i-1,j}^2 F_{i+1,ij}^2 \cdots F_{N,ij}^2 \tag{2.27}$$

$$\times \left(1 - \sum_{k=1, k\neq i,j}^{N} (1 - f(|\mathbf{x}_j - \mathbf{x}_k|)^2)\right)\left(1 - \sum_{k=1, k\neq i}^{N} (1 - f(|\mathbf{x}_i - \mathbf{x}_k|)^2)\right).$$

We now consider the first two terms on the right side of (2.23). In the numerator of the first term we use, for each fixed i, the estimate

$$f'(t_i)^2 \le \sum_{j=1}^{i-1} f'(\mathbf{x}_i - \mathbf{x}_j)^2 \,, \tag{2.28}$$

and in the second term we use $F_i \le f(|\mathbf{x}_i - \mathbf{x}_j|)$. For fixed i and j we can eliminate $\mathbf{x}_i$ and $\mathbf{x}_j$ from the rest of the integrand by using the bound (2.26) in the numerator and (2.27) in the denominator to do the $\mathbf{x}_i$ and $\mathbf{x}_j$ integrations. Note that, using partial integration, (2.3) implies that

$$\int \left(\mu f'(|\mathbf{x}|)^2 + \tfrac{1}{2}v(|\mathbf{x}|)f(|\mathbf{x}|)^2\right) d\mathbf{x} \le 4\pi a(1 - a/b)f_0(b)^{-2}. \tag{2.29}$$

Moreover, $f_0(b) \ge 1 - a/b$. We thus obtain

$$\int \left(2\mu f'(\mathbf{x}_i - \mathbf{x}_j)^2 + v(\mathbf{x}_i - \mathbf{x}_j)f(\mathbf{x}_i - \mathbf{x}_j)^2\right) d\mathbf{x}_i d\mathbf{x}_j \le 8\pi a\mu L^3 \left[1 - a/b\right]^{-1} \,. \tag{2.30}$$

In the denominator, we estimate

$$\int \left(1 - \sum_{p=1, p\neq i}^{N} (1 - f(|\mathbf{x}_p - \mathbf{x}_i|)^2)\right) d\mathbf{x}_i \ge L^3 - (N-1)I \,, \tag{2.31}$$

where we set $I = \int (1 - f(|\mathbf{x}|)^2)d\mathbf{x}$. Using that $f(|\mathbf{x}|) \geq [1 - a/|\mathbf{x}|]_+$ (see Appendix C), we get that $I \leq (4\pi/3)ab^2$. The same factor comes from the $\mathbf{x}_j$-integration. The remaining factors are identical in numerator and denominator and hence we conclude that the first and second term in (2.23) are bounded above by

$$\sum_{i=1}^{N}(i-1)\frac{8\pi a\mu}{(1-a/b)(1-I\rho_1)^2} \leq \frac{N(N-1)}{L^3}4\pi a\mu\left(1+O\left(a/b\right)\right) . \qquad (2.32)$$

A similar argument is now applied to the third term of (2.23). We omit the details. The result is an upper bound

$$\frac{2}{3}N(N-1)(N-2)\frac{\mu K^2}{(L^3-(N-1)I)^2} , \qquad (2.33)$$

with K given by $K = \int f(|\mathbf{x}|)f'(|\mathbf{x}|)d\mathbf{x}$. Using again that $[1-a/|\mathbf{x}|]_+ \leq f(|\mathbf{x}|) \leq 1$ as well as partial integration, we can estimate $K \leq 4\pi ab(1+O(a/b))$. Hence (2.33) is, for bounded a/b, bounded by const. $N\mu a^2\rho_1/b$. Combining this estimate with (2.32) proves (2.14). $\qquad \square$

2.2 Lower Bound

It was explained previously in this chapter why the lower bound for the bosonic ground state energy of (2.1) is not easy to obtain. The three different length scales (2.12) for bosons will play a role in the proof below.

- The scattering length a.

- The mean particle distance $\rho^{-1/3}$.

- The 'uncertainty principle length' ℓ_c, defined by $\mu\ell_c^{-2} = e_0(\rho)$, i.e., $\ell_c \sim (\rho a)^{-1/2}$.

Our lower bound for $e_0(\rho)$ is as follows.

Theorem 2.3 (Lower bound in the thermodynamic limit). *For a positive potential v with finite range and Y small enough*

$$\frac{e_0(\rho)}{4\pi\mu\rho a} \geq (1 - CY^{1/17}) \qquad (2.34)$$

with C a constant. If v does not have finite range, but decreases faster than $1/r^3$ (more precisely, $\int_R^\infty v(r)r^2dr < \infty$ for some R) then an analogous bound to (2.34) holds, but with $CY^{1/17}$ replaced by $o(1)$ as $Y \to 0$.

It should be noted right away that the error term $-C\,Y^{1/17}$ in (2.34) is of no fundamental significance and is not believed to reflect the true state of affairs. Presumably, it does not even have the right sign. We mention in passing that C can be taken to be 8.9 [Se1].

As mentioned at the beginning of this chapter after Eq. (2.2), a lower bound on $E_0(N, L)$ for finite N and L is of importance for applications to inhomogeneous gases, and in fact we derive (2.34) from such a bound. We state it in the following way:

Theorem 2.4 (Lower bound in a finite box). *For a positive potential v with finite range there is a $\delta > 0$ such that the ground state energy of (2.1) with Neumann boundary conditions satisfies*

$$E_0(N, L)/N \geq 4\pi\mu\rho a \left(1 - C\,Y^{1/17}\right) \tag{2.35}$$

for all N and L with $Y < \delta$ and $L/a > C'Y^{-6/17}$. Here C and C' are positive constants, independent of N and L. (Note that the condition on L/a requires in particular that N must be large enough, $N > \mathrm{const.}\,Y^{-1/17}$.) As in Theorem 2.3 such a bound, but possibly with a different error term, holds also for potentials v of infinite range that decrease sufficiently fast at infinity.

The first step in the proof of Theorem 2.4 is a generalization of a lemma of Dyson, which allows us to replace v by a 'soft' potential, at the cost of sacrificing kinetic energy and increasing the effective range.

Lemma 2.5. *Let $v(r) \geq 0$ with finite range R_0. Let $U(r) \geq 0$ be any function satisfying $\int U(r)r^2 dr \leq 1$ and $U(r) = 0$ for $r < R_0$. Let $\mathcal{B} \subset \mathbb{R}^3$ be star shaped with respect to 0 (e.g. convex with $0 \in \mathcal{B}$). Then for all differentiable functions ψ*

$$\int_{\mathcal{B}} \left[\mu|\nabla\psi|^2 + \tfrac{1}{2}v|\psi|^2\right] \geq \mu a \int_{\mathcal{B}} U|\psi|^2. \tag{2.36}$$

Proof. Actually, (2.36) holds with $\mu|\nabla\psi(\mathbf{x})|^2$ replaced by the (smaller) radial kinetic energy, $\mu|\partial\psi(\mathbf{x})/\partial r|^2$, and it suffices to prove the analog of (2.36) for the integral along each radial line with fixed angular variables. Along such a line we write $\psi(\mathbf{x}) = u(r)/r$ with $u(0) = 0$. We consider first the special case when U is a delta-function at some radius $R \geq R_0$, i.e.,

$$U(r) = \frac{1}{R^2}\delta(r - R). \tag{2.37}$$

For such U the analog of (2.36) along the radial line is

$$\int_0^{R_1} \{\mu[u'(r)-(u(r)/r)]^2+\tfrac{1}{2}v(r)|u(r)|^2\}dr \geq \begin{cases} 0 & \text{if } R_1 < R \\ \mu a|u(R)|^2/R^2 & \text{if } R \leq R_1 \end{cases} \qquad (2.38)$$

where R_1 is the length of the radial line segment in $\mathcal{B}$. The case $R_1 < R$ is trivial, because $\mu|\partial\psi/\partial r|^2 + \tfrac{1}{2}v|\psi|^2 \geq 0$. (Note that positivity of v is used here.) If $R \leq R_1$ we consider the integral on the left side of (2.38) from 0 to R instead of R_1 and minimize it under the boundary condition that $u(0) = 0$ and $u(R)$ is a fixed constant. Since everything is homogeneous in u we may normalize this value to $u(R) = R - a$. This minimization problem leads to the zero energy scattering equation (2.6). Since v is positive, the solution is a true minimum and not just a stationary point.

Because $v(r) = 0$ for $r > R_0$ the solution, u_0, satisfies $u_0(r) = r - a$ for $r > R_0$. By partial integration,

$$\int_0^R \{\mu[u_0'(r) - (u_0(r)/r)]^2 + \tfrac{1}{2}v(r)|u_0(r)|^2\}dr = \mu a|R - a|/R \geq \mu a|R - a|^2/R^2. \qquad (2.39)$$

But $|R - a|^2/R^2$ is precisely the right side of (2.38) if u satisfies the normalization condition.

This derivation of (2.36) for the special case (2.37) implies the general case, because every U can be written as a superposition of δ-functions, i.e., $U(r) = \int R^{-2}\delta(r - R)\,U(R)R^2dR$, and $\int U(R)R^2dR \leq 1$ by assumption. $\qquad\square$

By dividing Λ for given points $\mathbf{x}_1,\ldots,\mathbf{x}_N$ into Voronoi cells $\mathcal{B}_i$ that contain all points closer to $\mathbf{x}_i$ than to $\mathbf{x}_j$ with $j \neq i$ (these cells are star shaped w.r.t. $\mathbf{x}_i$, indeed convex), the following corollary of Lemma 2.5 can be derived in the same way as the corresponding Eq. (28) in [D1].

Corollary 2.6. *For any U as in Lemma 2.5*

$$H_N \geq \mu a W \qquad (2.40)$$

with W the multiplication operator

$$W(\mathbf{x}_1,\ldots,\mathbf{x}_N) = \sum_{i=1}^N U(t_i), \qquad (2.41)$$

where t_i is the distance of $\mathbf{x}_i$ to its nearest neighbor among the other points $\mathbf{x}_j$, $j = 1,\ldots,N$, i.e.,

$$t_i(\mathbf{x}_1,\ldots,\mathbf{x}_N) = \min_{j,\,j\neq i} |\mathbf{x}_i - \mathbf{x}_j|. \qquad (2.42)$$

(Note that t_i has here a slightly different meaning than in (2.17), where it denoted the distance to the nearest neighbor among the $\mathbf{x}_j$ with $j \leq i - 1$.)

Dyson considers in [D1] a one parameter family of U's that is essentially the same as the following choice, which is convenient for the present purpose:

$$U_R(r) = \begin{cases} 3(R^3 - R_0^3)^{-1} & \text{for } R_0 < r < R \\ 0 & \text{otherwise.} \end{cases} \tag{2.43}$$

We denote the corresponding interaction (2.41) by W_R. For the hard core gas one obtains

$$E(N, L) \geq \sup_R \; \inf_{(\mathbf{x}_1, \ldots, \mathbf{x}_N)} \mu a W_R(\mathbf{x}_1, \ldots, \mathbf{x}_N) \tag{2.44}$$

where the infimum is over $(\mathbf{x}_1, \ldots, x_N) \in \Lambda^N$ with $|\mathbf{x}_i - \mathbf{x}_j| \geq R_0 = a$, because of the hard core. At fixed R simple geometry gives

$$\inf_{(\mathbf{x}_1, \ldots, \mathbf{x}_N)} W_R(\mathbf{x}_1, \ldots, \mathbf{x}_N) \geq \left(\frac{A}{R^3} - \frac{B}{\rho R^6} \right) \tag{2.45}$$

with certain constants A and B. An evaluation of these constants gives Dyson's bound

$$E(N, L)/N \geq \frac{1}{10\sqrt{2}} 4\pi \mu \rho a. \tag{2.46}$$

The main reason this method does not give a better bound is that R must be chosen quite big, namely of the order of the mean particle distance $\rho^{-1/3}$, in order to guarantee that the spheres of radius R around the N points overlap. Otherwise the infimum of W_R will be zero. But large R means that W_R is small. It should also be noted that this method does not work for potentials other than hard spheres: If $|\mathbf{x}_i - \mathbf{x}_j|$ is allowed to be less than R_0, then the right side of (2.44) is zero because $U(r) = 0$ for $r < R_0$.

For these reasons we take another route. We still use Lemma 2.5 to get into the soft potential regime, but we do *not* sacrifice *all* the kinetic energy as in (2.40). Instead we write, for $\varepsilon > 0$

$$H_N = \varepsilon H_N + (1 - \varepsilon) H_N \geq \varepsilon T_N + (1 - \varepsilon) H_N \tag{2.47}$$

with $T_N = -\mu \sum_i \Delta_i$ and use (2.40) only for the part $(1 - \varepsilon) H_N$. This gives

$$H_N \geq \varepsilon T_N + (1 - \varepsilon) \mu a W_R. \tag{2.48}$$

We consider the operator on the right side from the viewpoint of first order perturbation theory, with εT_N as the unperturbed part, denoted H_0.

The ground state of H_0 in a box of side length L is $\Psi_0(\mathbf{x}_1, \ldots, \mathbf{x}_N) \equiv L^{-3N/2}$ and we denote expectation values in this state by $\langle \cdot \rangle_0$. A computation, cf. Eq. (21) in [LY1] (see also Eqs. (3.15)–(3.20)), gives

$$4\pi\rho \left(1 - \tfrac{1}{N}\right) \geq \langle W_R \rangle_0 / N$$
$$\geq 4\pi\rho \left(1 - \tfrac{1}{N}\right) \left(1 - \tfrac{2R}{L}\right)^3 \left(1 + 4\pi\rho(R^3 - R_0^3)/3\right)^{-1}. \quad (2.49)$$

The rationale behind the various factors is as follows: $(1 - \tfrac{1}{N})$ comes from the fact that the number of pairs is $N(N-1)/2$ and not $N^2/2$, $(1 - 2R/L)^3$ takes into account the fact that the particles do not interact beyond the boundary of Λ, and the last factor measures the probability to find another particle within the interaction range of the potential U_R for a given particle.

The estimates (2.49) on the first order term look at first sight quite promising, for if we let $L \to \infty$, $N \to \infty$ with $\rho = N/L^3$ fixed, and subsequently take $R \to 0$, then $\langle W_R \rangle_0 / N$ converges to $4\pi\rho$, which is just what is desired. But the first order result (2.49) is not a rigorous bound on $E_0(N, L)$, we need *error estimates*, and these will depend on ε, R and L.

We now recall *Temple's inequality* [T] for the expectation values of an operator $H = H_0 + V$ in the ground state $\langle \cdot \rangle_0$ of H_0. It is a simple consequence of the operator inequality

$$(H - E_0)(H - E_1) \geq 0 \quad (2.50)$$

for the two lowest eigenvalues, $E_0 < E_1$, of H and reads

$$E_0 \geq \langle H \rangle_0 - \frac{\langle H^2 \rangle_0 - \langle H \rangle_0^2}{E_1 - \langle H \rangle_0} \quad (2.51)$$

provided $E_1 - \langle H \rangle_0 > 0$. Furthermore, if $V \geq 0$ we may use $E_1 \geq E_1^{(0)} =$ second lowest eigenvalue of H_0 and replace E_1 in (2.51) by $E_1^{(0)}$.

From (2.49) and (2.51) we get the estimate

$$\frac{E_0(N, L)}{N} \geq 4\pi\mu a\rho \left(1 - \mathcal{E}(\rho, L, R, \varepsilon)\right) \quad (2.52)$$

with

$$1 - \mathcal{E}(\rho, L, R, \varepsilon)$$
$$= (1 - \varepsilon) \left(1 - \tfrac{1}{\rho L^3}\right) \left(1 - \tfrac{2R}{L}\right)^3 \left(1 + \tfrac{4\pi}{3}\rho(R^3 - R_0^3)\right)^{-1}$$
$$\times \left(1 - \frac{\mu a (\langle W_R^2 \rangle_0 - \langle W_R \rangle_0^2)}{\langle W_R \rangle_0 (E_1^{(0)} - \mu a \langle W_R \rangle_0)}\right). \quad (2.53)$$

To evaluate this further one may use the estimates (2.49) and the bound

$$\langle W_R^2 \rangle_0 \leq 3 \frac{N}{R^3 - R_0^3} \langle W_R \rangle_0 \tag{2.54}$$

which follows from $U_R^2 = 3(R^3 - R_0^3)^{-1} U_R$ together with the Schwarz inequality. A glance at the form of the error term reveals, however, that it is *not* possible here to take the thermodynamic limit $L \to \infty$ with ρ fixed: We have $E_1^{(0)} = \varepsilon \pi^2 \mu / L^2$ (this is the kinetic energy of a *single* particle in the first excited state in the box), and the factor $E_1^{(0)} - \mu a \langle W_R \rangle_0$ in the denominator in (2.53) is, up to unimportant constants and lower order terms, $\sim (\varepsilon L^{-2} - a\rho^2 L^3)$. Hence the denominator eventually becomes negative and Temple's inequality looses its validity if L is large enough.

As a way out of this dilemma we divide the big box Λ into cubic *cells* of side length ℓ that is kept *fixed* as $L \to \infty$. The number of cells, L^3/ℓ^3, on the other hand, increases with L. The N particles are distributed among these cells, and we use (2.53), with L replaced by ℓ, N by the particle number, n, in a cell and ρ by n/ℓ^3, to estimate the energy in each cell with *Neumann* conditions on the boundary. For each distribution of the particles we add the contributions from the cells, neglecting interactions across boundaries. Since $v \geq 0$ by assumption, this can only lower the energy. Finally, we minimize over all possible choices of the particle numbers for the various cells adding up to N. The energy obtained in this way is a lower bound to $E_0(N, L)$, because we are effectively allowing discontinuous test functions for the quadratic form given by H_N.

In mathematical terms, the cell method leads to

$$E_0(N, L)/N \geq (\rho \ell^3)^{-1} \inf \sum_{n \geq 0} c_n E_0(n, \ell) \tag{2.55}$$

where the infimum is over all choices of coefficients $c_n \geq 0$ (relative number of cells containing exactly n particles), satisfying the constraints

$$\sum_{n \geq 0} c_n = 1, \qquad \sum_{n \geq 0} c_n n = \rho \ell^3. \tag{2.56}$$

The minimization problem for the distributions of the particles among the cells would be easy if we knew that the ground state energy $E_0(n, \ell)$ (or a good lower bound to it) were convex in n. Then we could immediately conclude that it is best to have the particles as evenly distributed among the boxes as possible, i.e., c_n would be zero except for the n equal to the integer closest to $\rho \ell^3$. This would give

$$\frac{E_0(N, L)}{N} \geq 4\pi \mu a \rho \left(1 - \mathcal{E}(\rho, \ell, R, \varepsilon)\right) \tag{2.57}$$

i.e., replacement of L in (2.52) by ℓ, which is independent of L. The blow up of $\mathcal{E}$ for $L \to \infty$ would thus be avoided.

Since convexity of $E_0(n, \ell)$ is not known (except in the thermodynamic limit) we must resort to other means to show that $n = O(\rho \ell^3)$ in all boxes. The rescue comes from *superadditivity* of $E_0(n, \ell)$, i.e., the property

$$E_0(n + n', \ell) \geq E_0(n, \ell) + E_0(n', \ell) \tag{2.58}$$

which follows immediately from $v \geq 0$ by dropping the interactions between the n particles and the n' particles. The bound (2.58) implies in particular that for any $n, p \in \mathbb{N}$ with $n \geq p$

$$E_0(n, \ell) \geq [n/p] \, E_0(p, \ell) \geq \frac{n}{2p} E_0(p, \ell) \tag{2.59}$$

since the largest integer $[n/p]$ smaller than n/p is in any case $\geq n/(2p)$.

The way (2.59) is used is as follows: Replacing L by ℓ, N by n and ρ by n/ℓ^3 in (2.52) we have for fixed R and ε

$$E_0(n, \ell) \geq \frac{4\pi \mu a}{\ell^3} n(n-1) K(n, \ell) \tag{2.60}$$

with a certain function $K(n, \ell)$ determined by (2.53). We shall see that K is monotonously decreasing in n, so that if $p \in \mathbb{N}$ and $n \leq p$ then

$$E_0(n, \ell) \geq \frac{4\pi \mu a}{\ell^3} n(n-1) K(p, \ell). \tag{2.61}$$

We now split the sum in (2.55) into two parts. For $n < p$ we use (2.61), and for $n \geq p$ we use (2.59) together with (2.61) for $n = p$. The task is thus to minimize

$$\sum_{n<p} c_n n(n-1) + \frac{1}{2} \sum_{n \geq p} c_n n(p-1) \tag{2.62}$$

subject to the constraints (2.56). Putting

$$k := \rho \ell^3 \quad \text{and} \quad t := \sum_{n<p} c_n n \leq k \tag{2.63}$$

we have $\sum_{n \geq p} c_n n = k - t$, and since $n(n-1)$ is convex in n and vanishes for $n = 0$, and $\sum_{n<p} c_n \leq 1$, the expression (2.62) is

$$\geq t(t-1) + \frac{1}{2}(k-t)(p-1). \tag{2.64}$$

We have to minimize this for $1 \leq t \leq k$. If $p \geq 4k$ the minimum is taken at $t = k$ and is equal to $k(k-1)$. Altogether we have thus shown that

$$\frac{E_0(N, L)}{N} \geq 4\pi \mu a \rho \left(1 - \frac{1}{\rho \ell^3}\right) K(4\rho \ell^3, \ell). \tag{2.65}$$

What remains is to take a closer look at $K(4\rho\ell^3, \ell)$, which depends on the parameters ε and R besides ℓ, and choose the parameters in an optimal way. From (2.53) and (2.54) we obtain

$$K(n, \ell) = (1 - \varepsilon)\left(1 - \tfrac{2R}{\ell}\right)^3 \left(1 + \tfrac{4\pi}{3}(R^3 - R_0^3)\right)^{-1}$$
$$\times \left(1 - \frac{3}{\pi}\frac{an}{(R^3 - R_0^3)(\pi\varepsilon\ell^{-2} - 4a\ell^{-3}n(n-1))}\right). \tag{2.66}$$

The estimate (2.60) with this K is valid as long as the denominator in the last factor in (2.66) is ≥ 0, and in order to have a formula for all n we can take 0 as a trivial lower bound in other cases or when (2.60) is negative. As required for (2.61), K is monotonously decreasing in n. We now insert $n = 4\rho\ell^3$ and obtain

$$K(4\rho\ell^3, \ell) \geq (1 - \varepsilon)\left(1 - \tfrac{2R}{\ell}\right)^3 \left(1 + \text{const.}\right)Y(\ell/a)^3(R^3 - R_0^3)/\ell^3\right)^{-1}$$
$$\times \left(1 - \frac{\ell^3}{(R^3 - R_0^3)}\frac{\text{const.}\,Y}{(\varepsilon(a/\ell)^2 - (\text{const.})Y^2(\ell/a)^3)}\right) \tag{2.67}$$

with $Y = 4\pi\rho a^3/3$ as before. Also, the factor

$$\left(1 - \frac{1}{\rho\ell^3}\right) = (1 - (\text{const.})Y^{-1}(a/\ell)^3) \tag{2.68}$$

in (2.65) (which is the ratio between $n(n-1)$ and n^2) must not be forgotten. We now make the ansatz

$$\varepsilon \sim Y^\alpha, \quad a/\ell \sim Y^\beta, \quad (R^3 - R_0^3)/\ell^3 \sim Y^\gamma \tag{2.69}$$

with exponents α, β and γ that we choose in an optimal way. The conditions to be met are as follows:

- $\varepsilon(a/\ell)^2 - (\text{const.})Y^2(\ell/a)^3 > 0$. This holds for all small enough Y, provided $\alpha + 5\beta < 2$ which follows from the conditions below.

- $\alpha > 0$ in order that $\varepsilon \to 0$ for $Y \to 0$.

- $3\beta - 1 > 0$ in order that $Y^{-1}(a/\ell)^3 \to 0$ for $Y \to 0$.

- $1 - 3\beta + \gamma > 0$ in order that $Y(\ell/a)^3(R^3 - R_0^3)/\ell^3 \to 0$ for $Y \to 0$.

- $1 - \alpha - 2\beta - \gamma > 0$ to control the last factor in (2.67).

Taking

$$\alpha = 1/17, \quad \beta = 6/17, \quad \gamma = 3/17 \tag{2.70}$$

all these conditions are satisfied, and

$$\alpha = 3\beta - 1 = 1 - 3\beta + \gamma = 1 - \alpha - 2\beta - \gamma = 1/17. \tag{2.71}$$

It is also clear that $2R/\ell \sim Y^{\gamma/3} = Y^{1/17}$, up to higher order terms. This completes the proof of Theorems 2.3 and 2.4, for the case of potentials with finite range. By optimizing the proportionality constants in (2.69) one can show that $C = 8.9$ is possible in Theorem 2.3 [Se1]. The extension to potentials of infinite range but finite scattering length is obtained by approximation by finite range potentials, controlling the change of the scattering length as the cut-off is removed. See Appendix C for details. We remark that a slower decrease of the potential than $1/r^3$ implies infinite scattering length.   □

The exponents (2.70) mean in particular that

$$a \ll R \ll \rho^{-1/3} \ll \ell \ll (\rho a)^{-1/2}, \tag{2.72}$$

whereas Dyson's method required $R \sim \rho^{-1/3}$ as already explained. The condition $\rho^{-1/3} \ll \ell$ is required in order to have many particles in each box and thus $n(n-1) \approx n^2$. The condition $\ell \ll (\rho a)^{-1/2}$ is necessary for a spectral gap $\gg e_0(\rho)$ in Temple's inequality. It is also clear that this choice of ℓ would lead to a far too big energy and no bound for $e_0(\rho)$ if we had chosen Dirichlet instead of Neumann boundary conditions for the cells. But with the latter the method works!

Chapter 3

The Dilute Bose Gas in 2D

In contrast to the three-dimensional theory, the two-dimensional Bose gas began to receive attention only relatively late. The first derivation of the correct asymptotic formula was, to our knowledge, done by Schick [S] for a gas of hard discs. He found

$$e(\rho) \approx 4\pi\mu\rho|\ln(\rho a^2)|^{-1}. \tag{3.1}$$

This was accomplished by an infinite summation of 'perturbation series' diagrams. Subsequently, a corrected modification of [S] was given in [HFM]. Positive temperature extensions were given in [Po] and in [FH]. All this work involved an analysis in momentum space, with the exception of a method due to one of us that works directly in configuration space [L2]. Ovchinnikov [O] derived (3.1) by using, basically, the method in [L2]. These derivations require several unproven assumptions and are not rigorous.

In two dimensions the scattering length a is defined using the zero energy scattering equation (2.3) but instead of $\psi(r) \approx 1 - a/r$ we now impose the asymptotic condition $\psi(r) \approx \ln(r/a)$. This is explained in Appendix C.

Note that in two dimensions the ground state energy could not possibly be $e_0(\rho) \approx 4\pi\mu\rho a$ as in three dimensions because that would be dimensionally wrong. Since $e_0(\rho)$ should essentially be proportional to ρ, there is apparently no room for an a dependence — which is ridiculous! It turns out that this dependence comes about in the $\ln(\rho a^2)$ factor.

One of the intriguing facts about (3.1) is that the energy for N particles is *not equal* to $N(N-1)/2$ times the energy for two particles in the low density limit — as is the case in three dimensions. The latter quantity, $E_0(2, L)$, is, asymptotically for large L, equal to $8\pi\mu L^{-2} \left[\ln(L^2/a^2)\right]^{-1}$. (This is seen in an analogous way as (2.13). The three-dimensional boundary condition $\psi_0(|\mathbf{x}| = R) = 1 - a/R$ is replaced by $\psi_0(|\mathbf{x}| = R) = \ln(R/a)$ and moreover it has to be taken into account that with this normalization $\|\psi_0\|^2 = (\text{volume})(\ln(R/a))^2$ (to leading order), instead of just the volume in the three-dimensional case.) Thus, if the $N(N-1)/2$ rule were to apply, (3.1) would have to be replaced by the much smaller quantity $4\pi\mu\rho \left[\ln(L^2/a^2)\right]^{-1}$. In other words, L, which tends to ∞ in the thermodynamic limit, has to be replaced by the mean particle separation, $\rho^{-1/2}$ in the logarithmic factor. Various poetic formulations of this curious fact have been given, but the fact remains that the non-linearity is something that does not occur in more than two dimensions and its precise nature is hardly obvious, physically. This anomaly is the main reason that the two-dimensional case is not a trivial extension of the three-dimensional one.

Eq. (3.1) was proved in [LY2] for nonnegative, finite range two-body potentials by finding upper and lower bounds of the correct form, using similar ideas as in the previous chapter for the three-dimensional case. We discuss below the modifications that have to be made in the present two-dimensional case. The restriction to finite range can be relaxed as in three dimensions, but the restriction to nonnegative v cannot be removed in the current state of our methodology. The upper bounds will have relative remainder terms $O(|\ln(\rho a^2)|^{-1})$ while the lower bound will have remainder $O(|\ln(\rho a^2)|^{-1/5})$. It is claimed in [HFM] that the relative error for a hard core gas is negative and $O(\ln|\ln(\rho a^2)||\ln(\rho a^2)|^{-1})$, which is consistent with our bounds.

The upper bound is derived in complete analogy with the three dimensional case. The function f_0 in the variational ansatz (2.19) is in two dimensions also the zero energy scattering solution — but for 2D, of course. The result is

$$E_0(N, L)/N \leq \frac{2\pi\mu\rho}{\ln(b/a) - \pi\rho b^2} \left(1 + O([\ln(b/a)]^{-1})\right). \qquad (3.2)$$

The minimum over b of the leading term is obtained for $b = (2\pi\rho)^{-1/2}$. Inserting this in (3.2) we thus obtain

$$E_0(N, L)/N \leq \frac{4\pi\mu\rho}{|\ln(\rho a^2)|} \left(1 + O(|\ln(\rho a^2)|^{-1})\right). \qquad (3.3)$$

To prove the lower bound the essential new step is to modify Dyson's lemma for 2D. The 2D version of Lemma 2.5 is:

Lemma 3.1. *Let $v(r) \geq 0$ and $v(r) = 0$ for $r > R_0$. Let $U(r) \geq 0$ be any function satisfying*

$$\int_0^\infty U(r) \ln(r/a) r \, dr \leq 1 \qquad \text{and} \qquad U(r) = 0 \quad \text{for } r < R_0. \qquad (3.4)$$

Let $\mathcal{B} \subset \mathbb{R}^2$ be star-shaped with respect to 0 (e.g. convex with $0 \in \mathcal{B}$). Then, for all functions ψ in the Sobolev space $H^1(\mathcal{B})$,

$$\int_{\mathcal{B}} \left(\mu |\nabla \psi(\mathbf{x})|^2 + \tfrac{1}{2} v(|\mathbf{x}|) |\psi(\mathbf{x})|^2 \right) \, d\mathbf{x} \geq \mu \int_{\mathcal{B}} U(|\mathbf{x}|) |\psi(\mathbf{x})|^2 \, d\mathbf{x}. \qquad (3.5)$$

Proof. In polar coordinates, r, θ, one has $|\nabla \psi|^2 \geq |\partial \psi / \partial r|^2$. Therefore, it suffices to prove that for each angle $\theta \in [0, 2\pi)$, and with $\psi(r, \theta)$ denoted simply by $f(r)$,

$$\int_0^{R(\theta)} \left(\mu |\partial f(r)/\partial r|^2 + \tfrac{1}{2} v(r) |f(r)|^2 \right) r \, dr \geq \mu \int_0^{R(\theta)} U(r) |f(r)|^2 \, r \, dr, \qquad (3.6)$$

where $R(\theta)$ denotes the distance of the origin to the boundary of $\mathcal{B}$ along the ray θ.

If $R(\theta) \leq R_0$ then (3.6) is trivial because the right side is zero while the left side is evidently nonnegative. (Here, $v \geq 0$ is used.)

If $R(\theta) > R_0$ for some given value of θ, consider the disc $\mathcal{D}(\theta) = \{\mathbf{x} \in \mathbb{R}^2 \; 0 \leq |\mathbf{x}| \leq R(\theta)\}$ centered at the origin in $\mathbb{R}^2$ and of radius $R(\theta)$. Our function f defines a radial function, $\mathbf{x} \mapsto f(|\mathbf{x}|)$ on $\mathcal{D}(\theta)$, and (3.6) is equivalent to

$$\int_{\mathcal{D}(\theta)} \left(\mu |\nabla f(|\mathbf{x}|)|^2 + \frac{1}{2} v(|\mathbf{x}|) |f(|\mathbf{x}|)|^2 \right) d\mathbf{x} \geq \mu \int_{\mathcal{D}(\theta)} U(|\mathbf{x}|) |f(|\mathbf{x}|)|^2 d\mathbf{x}. \qquad (3.7)$$

Now choose some $R \in (R_0, \; R(\theta))$ and note that the left side of (3.7) is not smaller than the same quantity with $\mathcal{D}(\theta)$ replaced by the smaller disc $\mathcal{D}_R = \{\mathbf{x} \in \mathbb{R}^2 \; 0 \leq |\mathbf{x}| \leq R\}$. (Again, $v \geq 0$ is used.) We now minimize this integral over $\mathcal{D}_R$, fixing $f(R)$. This minimization problem leads to the zero energy scattering equation. Plugging in the solution and integrating by parts leads to

$$2\pi \int_0^{R(\theta)} \left(\mu |\partial f(r)/\partial r|^2 + \frac{1}{2} v(r) |f(r)|^2 \right) r \, dr \geq \frac{2\pi \mu}{\ln(R/a)} |f(R)|^2. \qquad (3.8)$$

The proof is completed by multiplying both sides of (3.8) by $U(R) R \ln(R/a)$ and integrating with respect to R from R_0 to $R(\theta)$. $\qquad \square$

As in Corollary 2.6, Lemma 3.1 can be used to bound the many body Hamiltonian H_N from below, as follows:

Corollary 3.2. *For any U as in Lemma 3.1 and any $0 < \varepsilon < 1$*

$$H_N \geq \varepsilon T_N + (1 - \varepsilon) \mu W \qquad (3.9)$$

with $T_N = -\mu \sum_{i=1}^N \Delta_i$ and

$$W(\mathbf{x}_1, \dots, \mathbf{x}_N) = \sum_{i=1}^N U \left(\min_{j, \, j \neq i} |\mathbf{x}_i - \mathbf{x}_j| \right). \qquad (3.10)$$

For U we choose the following functions, parameterized by $R > R_0$:

$$U_R(r) = \begin{cases} \nu(R)^{-1} & \text{for } R_0 < r < R \\ 0 & \text{otherwise} \end{cases} \tag{3.11}$$

with $\nu(R)$ chosen so that

$$\int_{R_0}^{R} U_R(r) \ln(r/a) r \, dr = 1 \tag{3.12}$$

for all $R > R_0$, i.e.,

$$\nu(R) = \int_{R_0}^{R} \ln(r/a) r \, dr = \tfrac{1}{4} \left\{ R^2 \left(\ln(R^2/a^2) - 1 \right) - R_0^2 \left(\ln(R_0^2/a^2) - 1 \right) \right\}. \tag{3.13}$$

The nearest neighbor interaction (3.10) corresponding to U_R will be denoted W_R.

As in Section 2.2 we shall need estimates on the expectation value, $\langle W_R \rangle_0$, of W_R in the ground state of εT_N of (3.9) with Neumann boundary conditions. This is just the average value of W_R in a hypercube in $\mathbb{R}^{2N}$. Besides the normalization factor $\nu(R)$, the computation involves the volume (area) of the support of U_R, which is

$$A(R) = \pi(R^2 - R_0^2). \tag{3.14}$$

In contrast to the three-dimensional situation the normalization factor $\nu(R)$ is not just a constant (R independent) multiple of $A(R)$; the factor $\ln(r/a)$ in (3.4) accounts for the more complicated expressions in the two-dimensional case. Taking into account that U_R is proportional to the characteristic function of a disc of radius R with a hole of radius R_0, the following inequalities for n particles in a box of side length ℓ are obtained by the same geometric reasoning as lead to (2.49), cf. [LY1]:

$$\langle W_R \rangle_0 \;\geq\; \frac{n}{\nu(R)} \left(1 - \tfrac{2R}{\ell}\right)^2 \left[1 - (1-Q)^{(n-1)}\right] \tag{3.15}$$

$$\langle W_R \rangle_0 \;\leq\; \frac{n}{\nu(R)} \left[1 - (1-Q)^{(n-1)}\right] \tag{3.16}$$

with

$$Q = A(R)/\ell^2 \tag{3.17}$$

being the relative volume occupied by the support of the potential U_R. Since $U_R^2 = \nu(R)^{-1} U_R$ we also have

$$\langle W_R^2 \rangle_0 \leq \frac{n}{\nu(R)} \langle W_R \rangle_0. \tag{3.18}$$

As in [LY1] we estimate $[1 - (1 - Q)^{(n-1)}]$ by

$$(n - 1)Q \geq \left[1 - (1 - Q)^{(n-1)}\right] \geq \frac{(n-1)Q}{1 + (n-1)Q} \tag{3.19}$$

This gives

$$\langle W_R \rangle_0 \geq \frac{n(n-1)}{\nu(R)} \cdot \frac{Q}{1 + (n-1)Q}, \tag{3.20}$$

$$\langle W_R \rangle_0 \leq \frac{n(n-1)}{\nu(R)} \cdot Q \, . \tag{3.21}$$

From Temple's inequality [T] we obtain like in (2.51) the estimate

$$E_0(n, \ell) \geq (1 - \varepsilon)\langle W_R \rangle_0 \left(1 - \frac{\mu(\langle W_R^2 \rangle_0 - \langle W_R \rangle_0^2)}{\langle W_R \rangle_0 (E_1^{(0)} - \mu\langle W_R \rangle_0)}\right) \tag{3.22}$$

where

$$E_1^{(0)} = \frac{\varepsilon\mu}{\ell^2} \tag{3.23}$$

is the energy of the lowest excited state of εT_n. This estimate is valid for $E_1^{(0)}/\mu > \langle W_R \rangle_0$, i.e., it is important that ℓ is not too big.

Putting (3.20)–(3.22) together we obtain the estimate

$$E_0(n, \ell) \geq \frac{n(n-1)}{\ell^2} \frac{A(R)}{\nu(R)} K(n) \tag{3.24}$$

with

$$K(n) = (1 - \varepsilon) \cdot \frac{(1 - \frac{2R}{\ell})^2}{1 + (n-1)Q} \cdot \left(1 - \frac{n}{(\varepsilon\nu(R)/\ell^2) - n(n-1)Q}\right) \tag{3.25}$$

Note that Q depends on ℓ and R, and K depends on ℓ, R and ε besides n. We have here dropped the term $\langle W_R \rangle_0^2$ in the numerator in (3.22), which is appropriate for the purpose of a lower bound.

We note that K is monotonically decreasing in n, so for a given n we may replace $K(n)$ by $K(p)$ provided $p \geq n$. As explained in the previous chapter, (2.58)–(2.65), convexity of $n \mapsto n(n-1)$ together with superadditivity of $E_0(n, \ell)$ in n leads, for $p = 4\rho\ell^2$, to an estimate for the energy of N particles in the large box when the side length L is an integer multiple of ℓ:

$$E_0(N, L)/N \geq \frac{\rho A(R)}{\nu(R)} \left(1 - \frac{1}{\rho\ell^2}\right) K(4\rho\ell^2) \tag{3.26}$$

with $\rho = N/L^2$.

Let us now look at the conditions on the parameters ε, R and ℓ that have to be met in order to obtain a lower bound with the same leading term as the upper bound (3.3).

From (3.13) we have

$$\frac{A(R)}{\nu(R)} = \frac{4\pi}{(\ln(R^2/a^2) - 1)} \left(1 - O((R_0^2/R^2) \ln(R/R_0))\right) \tag{3.27}$$

We thus see that as long as $a < R < \rho^{-1/2}$ the logarithmic factor in the denominator in (3.27) has the right form for a lower bound. Moreover, for Temple's inequality the denominator in the third factor in (3.25) must be positive. With $n = 4\rho\ell^2$ and $\nu(R) \geq$ (const.)$R^2 \ln(R^2/a^2)$ for $R \gg R_0$, this condition amounts to

$$(\text{const.})\varepsilon \ln(R^2/a^2)/\ell^2 > \rho^2\ell^4. \tag{3.28}$$

The relative error terms in (3.26) that have to be $\ll 1$ are

$$\varepsilon, \quad \frac{1}{\rho\ell^2}, \quad \frac{R}{\ell}, \quad \rho R^2, \quad \frac{\rho\ell^4}{\varepsilon R^2 \ln(R^2/a^2)}. \tag{3.29}$$

We now choose

$$\varepsilon \sim |\ln(\rho a^2)|^{-1/5}, \quad \ell \sim \rho^{-1/2}|\ln(\rho a^2)|^{1/10}, \quad R \sim \rho^{-1/2}|\ln(\rho a^2)|^{-1/10} \tag{3.30}$$

Condition (3.28) is satisfied since the left side is $>$ (const.)$|\ln(\rho a^2)|^{3/5}$ and the right side is $\sim |\ln(\rho a^2)|^{2/5}$. The first three error terms in (3.29) are all of the same order, $|\ln(\rho a^2)|^{-1/5}$, the last is $\sim |\ln(\rho a^2)|^{-1/5}(\ln|\ln(\rho a^2)|)^{-1}$. With these choices, (3.26) thus leads to the following:

Theorem 3.3 (Lower bound). *For all N and L large enough, such that $L >$ (const.)$\rho^{-1/2}|\ln(\rho a^2)|^{1/10}$ and $N >$ (const.)$|\ln(\rho a^2)|^{1/5}$ with $\rho = N/L^2$, the ground state energy with Neumann boundary condition satisfies*

$$E_0(N,L)/N \geq \frac{4\pi\mu\rho}{|\ln(\rho a^2)|} \left(1 - O(|\ln(\rho a^2)|^{-1/5})\right). \tag{3.31}$$

In combination with the upper bound (3.3) this also proves

Theorem 3.4 (Energy at low density in the thermodynamic limit).

$$\lim_{\rho a^2 \to 0} \frac{e_0(\rho)}{4\pi\mu\rho|\ln(\rho a^2)|^{-1}} = 1 \tag{3.32}$$

where $e_0(\rho) = \lim_{N\to\infty} E_0(N, \rho^{-1/2}N^{1/2})/N$. This holds irrespective of boundary conditions.

As in the three-dimensional case, Theorem 3.4 is also valid for an infinite range potential v provided that $v \geq 0$ and for some R we have $\int_R^\infty v(r)r\, dr < \infty$, which guarantees a finite scattering length.

Chapter 4

Generalized Poincaré Inequalities

This chapter contains some lemmas that are of independent mathematical interest, but whose significance for the physics of the Bose gas may not be obvious at this point. They will, however, turn out to be important tools for the discussion of Bose-Einstein condensation (BEC) and superfluidity in the next chapters.

The classic Poincaré inequality [LLo] bounds the L^q-norm of a function, f, orthogonal to a given function g in a domain $\mathcal{K}$, in terms of some L^p-norm of its gradient in $\mathcal{K}$. For the proof of BEC we shall need a generalization of this inequality where the estimate is in terms of the gradient of f on a subset $\Omega \subset \mathcal{K}$ and a remainder that tends to zero with the volume of the complement $\Omega^c = \mathcal{K} \setminus \Omega$. For superfluidity it will be necessary to generalize this further by adding a vector potential to the gradient. This is the most complex of the lemmas because the other two can be derived directly from the classical Poincaré inequality using Hölder's inequality. The first lemma is the simplest variant and it is sufficient for the discussion of BEC in the case of a homogeneous gas. In this case the function g can be taken to be the constant function. The same holds for the second lemma, which will be used for the discussion of superfluidity in a homogeneous gas with periodic boundary conditions, but the modification of the gradient requires a more elaborate proof. The last lemma, that will be used for the discussion of BEC in the inhomogeneous case, is again a simple consequence of the classic Poincaré and Hölder inequalities. For a more comprehensive discussion of generalized Poincaré inequalities with further generalizations we refer to [LSeY7].

Lemma 4.1 (Generalized Poincaré inequality: Homogeneous case). *Let $\mathcal{K} \subset \mathbb{R}^3$ be a cube of side length L, and define the average of a function $f \in L^1(\mathcal{K})$ by*

$$\langle f \rangle_{\mathcal{K}} = \frac{1}{L^3} \int_{\mathcal{K}} f(\mathbf{x}) \, d\mathbf{x} \ .$$

There exists a constant C such that for all measurable sets $\Omega \subset \mathcal{K}$ and all $f \in H^1(\mathcal{K})$ the inequality

$$\int_{\mathcal{K}} |f(\mathbf{x}) - \langle f \rangle_{\mathcal{K}}|^2 d\mathbf{x} \leq C \left(L^2 \int_{\Omega} |\nabla f(\mathbf{x})|^2 d\mathbf{x} + |\Omega^c|^{2/3} \int_{\mathcal{K}} |\nabla f(\mathbf{x})|^2 d\mathbf{x} \right) \quad (4.1)$$

holds. Here $\Omega^c = \mathcal{K} \setminus \Omega$, and $|\cdot|$ denotes the measure of a set. .

Proof. By scaling, it suffices to consider the case $L = 1$. Using the usual Poincaré-Sobolev inequality on $\mathcal{K}$ (see [LLo], Thm. 8.12), we infer that there exists a $C > 0$ such that

$$
\begin{aligned}
\|f - \langle f \rangle_{\mathcal{K}}\|_{L^2(\mathcal{K})}^2 &\leq \tfrac{1}{2} C \|\nabla f\|_{L^{6/5}(\mathcal{K})}^2 \\
&\leq C \left(\|\nabla f\|_{L^{6/5}(\Omega)}^2 + \|\nabla f\|_{L^{6/5}(\Omega^c)}^2 \right) \ .
\end{aligned}
\quad (4.2)
$$

Applying Hölder's inequality

$$\|\nabla f\|_{L^{6/5}(\Omega)} \leq \|\nabla f\|_{L^2(\Omega)} |\Omega|^{1/3}$$

(and the analogue with Ω replaced by Ω^c), we see that (4.1) holds. $\square$

In the next lemma $\mathcal{K}$ is again a cube of side length L, but we now replace the gradient ∇ by

$$\nabla_{\varphi} := \nabla + i(0, 0, \varphi/L), \quad (4.3)$$

where φ is a real parameter, and require periodic boundary conditions on $\mathcal{K}$.

Lemma 4.2 (Generalized Poincaré inequality with a vector potential). *For any $|\varphi| < \pi$ there are constants $c > 0$ and $C < \infty$ such that for all subsets $\Omega \subset \mathcal{K}$ and all functions $f \in H^1(\mathcal{K})$ with periodic boundary conditions on $\mathcal{K}$ the following estimate holds:*

$$
\|\nabla_{\varphi} f\|_{L^2(\Omega)}^2 \geq \frac{\varphi^2}{L^2} \|f\|_{L^2(\mathcal{K})}^2 + \frac{c}{L^2} \|f - \langle f \rangle_{\mathcal{K}}\|_{L^2(\mathcal{K})}^2
$$
$$
- C \left(\|\nabla_{\varphi} f\|_{L^2(\mathcal{K})}^2 + \frac{1}{L^2} \|f\|_{L^2(\mathcal{K})}^2 \right) \left(\frac{|\Omega^c|}{L^3} \right)^{1/2} \ . \quad (4.4)
$$

Here $|\Omega^c|$ is the volume of $\Omega^c = \mathcal{K} \setminus \Omega$, the complement of Ω in $\mathcal{K}$.

Proof. We shall derive (4.4) from a special form of this inequality that holds for all functions that are orthogonal to the constant function. Namely, for any positive $\alpha < 2/3$ and some constants $c > 0$ and $\tilde{C} < \infty$ (depending only on α and $|\varphi| < \pi$) we claim that

$$\|\nabla_\varphi h\|^2_{L^2(\Omega)} \geq \frac{\varphi^2 + c}{L^2} \|h\|^2_{L^2(\mathcal{K})} - \tilde{C} \left(\frac{|\Omega^c|}{L^3} \right)^\alpha \|\nabla_\varphi h\|^2_{L^2(\mathcal{K})} , \tag{4.5}$$

provided $\langle 1, h \rangle_\mathcal{K} = 0$. (Remark: Eq. (4.5) holds also for $\alpha = 2/3$, but the proof is slightly more complicated in that case. See [LSeY7].) If (4.5) is known the derivation of (4.4) is easy: For any f, the function $h = f - L^{-3}\langle 1, f \rangle_\mathcal{K}$ is orthogonal to 1. Moreover,

$$\begin{aligned}
\|\nabla_\varphi h\|^2_{L^2(\Omega)} &= \|\nabla_\varphi h\|^2_{L^2(\mathcal{K})} - \|\nabla_\varphi h\|^2_{L^2(\Omega^c)} \\
&= \|\nabla_\varphi f\|^2_{L^2(\Omega)} - \frac{\varphi^2}{L^2} |\langle L^{-3/2}, f \rangle_\mathcal{K}|^2 \left(1 + \frac{|\Omega^c|}{L^3} \right) \\
&\quad + 2\frac{\varphi}{L} \mathrm{Re} \, \langle L^{-3/2}, f \rangle_\mathcal{K} \langle \nabla_\varphi f, L^{-3/2} \rangle_{\Omega^c} \\
&\leq \|\nabla_\varphi f\|^2_{L^2(\Omega)} - \frac{\varphi^2}{L^2} |\langle L^{-3/2}, f \rangle_\mathcal{K}|^2 \\
&\quad + \frac{|\varphi|}{L} \left(L\|\nabla_\varphi f\|^2_{L^2(\mathcal{K})} + \frac{1}{L}\|f\|^2_{L^2(\mathcal{K})} \right) \left(\frac{|\Omega^c|}{L^3} \right)^{1/2}
\end{aligned} \tag{4.6}$$

and

$$\begin{aligned}
\frac{\varphi^2 + c}{L^2} \|h\|^2_{L^2(\mathcal{K})} &= \frac{\varphi^2}{L^2} \left(\|f\|^2_{L^2(\mathcal{K})} - |\langle L^{-3/2}, f \rangle_\mathcal{K}|^2 \right) \\
&\quad + \frac{c}{L^2} \|f - L^{-3}\langle 1, f \rangle_\mathcal{K}\|^2_{L^2(\mathcal{K})} .
\end{aligned} \tag{4.7}$$

Setting $\alpha = \frac{1}{2}$, using $\|\nabla_\varphi h\|_{L^2(\mathcal{K})} \leq \|\nabla_\varphi f\|_{L^2(\mathcal{K})}$ in the last term in (4.5) and combining (4.5), (4.6) and (4.7) gives (4.4) with $C = |\varphi| + \tilde{C}$.

We now turn to the proof of (4.5). For simplicity we set $L = 1$. The general case follows by scaling. Assume that (4.5) is false. Then there exist sequences of constants $C_n \to \infty$, functions h_n with $\|h_n\|_{L^2(\mathcal{K})} = 1$ and $\langle 1, h_n \rangle_\mathcal{K} = 0$, and domains $\Omega_n \subset \mathcal{K}$ such that

$$\lim_{n \to \infty} \left\{ \|\nabla_\varphi h_n\|^2_{L^2(\Omega_n)} + C_n |\Omega_n^c|^\alpha \|\nabla_\varphi h_n\|^2_{L^2(\mathcal{K})} \right\} \leq \varphi^2 . \tag{4.8}$$

We shall show that this leads to a contradiction.

Since the sequence h_n is bounded in $L^2(\mathcal{K})$, it has a subsequence, deno-
ted again by h_n, that converges weakly to some $h \in L^2(\mathcal{K})$ (i.e., $\langle g, h_n \rangle_\mathcal{K} \to$
$\langle g, h \rangle_\mathcal{K}$ for all $g \in L^2(\mathcal{K})$). Moreover, by Hölder's inequality the $L^p(\Omega_n^c)$ norm
$\|\nabla_\varphi h_n\|_{L^p(\Omega_n^c)} = (\int_{\Omega_n^c} |\nabla_\varphi h(\mathbf{x})|^p d\mathbf{x})^{1/p}$ is bounded by $|\Omega_n^c|^{\alpha/2} \|\nabla_\varphi h_n\|_{L^2(\mathcal{K})}$ for
$p = 2/(\alpha + 1)$. From (4.8) we conclude that $\|\nabla_\varphi h_n\|_{L^p(\Omega_n^c)}$ is bounded and also
that $\|\nabla_\varphi h_n\|_{L^p(\Omega_n)} \leq \|\nabla_\varphi h_n\|_{L^2(\Omega_n)}$ is bounded. Altogether, $\nabla_\varphi h_n$ is bounded
in $L^p(\mathcal{K})$, and by passing to a further subsequence if necessary, we can therefore
assume that $\nabla_\varphi h_n$ converges weakly in $L^p(\mathcal{K})$. The same applies to ∇h_n. Since
$p' = 2/(\alpha + 1)$ with $\alpha < 2/3$ the hypotheses of the Rellich-Kondrashov Theorem
[LLo, Thm. 8.9] are fulfilled and consequently h_n converges *strongly* in $L^2(\mathcal{K})$ to
h (i.e., $\|h - h_n\|_{L^2(\mathcal{K})} \to 0$). We shall now show that

$$\liminf_{n \to \infty} \|\nabla_\varphi h_n\|_{L^2(\Omega_n)}^2 \geq \|\nabla_\varphi h\|_{L^2(\mathcal{K})}^2 . \tag{4.9}$$

This will complete the proof because the h_n are normalized and orthogonal to 1
and the same holds for h by strong convergence. Hence the right side of (4.9) is
necessarily $> \varphi^2$, since for $|\varphi| < \pi$ the lowest eigenvalue of $-\nabla_\varphi^2$, with constant
eigenfunction, is non-degenerate. This contradicts (4.8).

Eq. (4.9) is essentially a consequence of the weak lower semicontinuity of
the L^2 norm, but the dependence on Ω_n leads to a slight complication. First,
Eq. (4.8) and $C_n \to \infty$ clearly imply that $|\Omega_n^c| \to 0$, because $\|\nabla_\varphi h_n\|_{L^2(\mathcal{K})}^2 > \varphi^2$.
By choosing a subsequence we may assume that $\sum_n |\Omega_n^c| < \infty$. For some fixed
N let $\widetilde{\Omega}_N = \mathcal{K} \setminus \cup_{n \geq N} \Omega_n^c$. Then $\widetilde{\Omega}_N \subset \Omega_n$ for $n \geq N$. Since $\|\nabla_\varphi h_n\|_{L^2(\Omega_n)}^2$ is
bounded, $\nabla_\varphi h_n$ is also bounded in $L^2(\widetilde{\Omega}_N)$ and a subsequence of it converges
weakly in $L^2(\widetilde{\Omega}_N)$ to $\nabla_\varphi h$. Hence

$$\liminf_{n \to \infty} \|\nabla_\varphi h_n\|_{L^2(\Omega_n)}^2 \geq \liminf_{n \to \infty} \|\nabla_\varphi h_n\|_{L^2(\widetilde{\Omega}_N)}^2 \geq \|\nabla_\varphi h\|_{L^2(\widetilde{\Omega}_N)}^2 . \tag{4.10}$$

Since $\widetilde{\Omega}_N \subset \widetilde{\Omega}_{N+1}$ and $\cup_N \widetilde{\Omega}_N = \mathcal{K}$ (up to a set of measure zero), we can now let
$N \to \infty$ on the right side of (4.10). By monotone convergence this converges to
$\|\nabla_\varphi h\|_{L^2(\mathcal{K})}^2$. This proves (4.9) which, as remarked above, contradicts (4.8). $\square$

The last lemma is a simple generalization of Lemma 4.1 with $\mathcal{K} \subset \mathbb{R}^m$ a
bounded and connected set that is sufficiently nice so that the Poincaré-Sobolev
inequality (see [LLo, Thm. 8.12]) holds on $\mathcal{K}$. In particular, this is the case if
$\mathcal{K}$ satisfies the cone property [LLo] (e.g., if $\mathcal{K}$ is a rectangular box or a cube).
Moreover, the constant function on $\mathcal{K}$ is here replaced by a more general bounded
function.

Lemma 4.3 (Generalized Poincaré inequality: Inhomog. case). *For $d \geq 2$ let $\mathcal{K} \subset \mathbb{R}^d$ be as explained above, and let h be a bounded function with $\int_{\mathcal{K}} h = 1$. There exists a constant C (depending only on $\mathcal{K}$ and h) such that for all measurable sets $\Omega \subset \mathcal{K}$ and all $f \in H^1(\mathcal{K})$ with $\int_{\mathcal{K}} f h \, d\mathbf{x} = 0$, the inequality*

$$\int_{\mathcal{K}} |f(\mathbf{x})|^2 d\mathbf{x} \leq C \left(\int_{\Omega} |\nabla f(\mathbf{x})|^2 d\mathbf{x} + \left(\frac{|\Omega^c|}{|\mathcal{K}|} \right)^{2/d} \int_{\mathcal{K}} |\nabla f(\mathbf{x})|^2 d\mathbf{x} \right) \qquad (4.11)$$

holds. Here $|\cdot|$ denotes the measure of a set, and $\Omega^c = \mathcal{K} \setminus \Omega$.

Proof. By the usual Poincaré-Sobolev inequality on $\mathcal{K}$ (see [LLo, Thm. 8.12]),

$$
\begin{aligned}
\|f\|_{L^2(\mathcal{K})}^2 &\leq \tilde{C} \|\nabla f\|_{L^{2d/(d+2)}(\mathcal{K})}^2 \\
&\leq 2\tilde{C} \left(\|\nabla f\|_{L^{2d/(d+2)}(\Omega)}^2 + \|\nabla f\|_{L^{2d/(d+2)}(\Omega^c)}^2 \right),
\end{aligned}
\qquad (4.12)
$$

if $d \geq 2$ and $\int_{\mathcal{K}} f h = 0$. Applying Hölder's inequality

$$\|\nabla f\|_{L^{2d/(d+2)}(\Omega)} \leq \|\nabla f\|_{L^2(\Omega)} |\Omega|^{1/d}$$

(and the analogue with Ω replaced by Ω^c), we see that (4.4) holds with $C = 2|\mathcal{K}|^{2/d}\tilde{C}$. $\qquad \square$

Chapter 5

Bose-Einstein Condensation and Superfluidity for Homogeneous Gases

5.1 Bose-Einstein Condensation

Bose-Einstein condensation (BEC) is the phenomenon of a macroscopic occupation of a single one-particle quantum state, discovered by Einstein for thermal equilibrium states of an ideal Bose gas at sufficiently low temperatures [E]. We are here concerned with interacting Bose gases, where the question of the existence of BEC is highly nontrivial even for the ground state. Due to the interaction the many body ground state is not a product of one-particle states but the concept of a macroscopic occupation of a single state acquires a precise meaning through the *one-particle density matrix*, as discussed in Section 1.2. Namely, this is the statement that the operator on $L^2(\mathbb{R}^d)$ ($d = 2$ or 3) given by the kernel

$$\gamma(\mathbf{x}, \mathbf{x}') = N \int \Psi_0(\mathbf{x}, \mathbf{X}) \Psi_0(\mathbf{x}', \mathbf{X}) d\mathbf{X} \tag{5.1}$$

has an eigenvalue of order N. Here, Ψ_0 denotes the normalized ground state wave function. In case the eigenfunction corresponding to the largest eigenvalue of γ is constant (or, at least, not orthogonal to the constant function), this means that

$$\frac{1}{L^d} \int\int \gamma(\mathbf{x}, \mathbf{y}) d\mathbf{x} d\mathbf{y} \geq cN \tag{5.2}$$

for all large N, with $c > 0$ depending only on the density N/L^d.

The problem remains open after more than 75 years since the first investigations on the Bose gas [B, E]. Our construction in Chapter 2 shows that (in 3D) BEC exists on a length scale of order $\rho^{-1/3}Y^{-1/17}$ which, unfortunately, is not a 'thermodynamic' length like volume$^{1/3}$. The same remark applies to the 2D case of Chapter 3, where BEC is proved over a length scale $\rho^{-1/10}|\ln(\rho a^2)|^{1/10}$.

In a certain limit, however, one can prove (5.2), as has been shown in [LSe]. In this limit the interaction potential v is varied with N so that the ratio a/L of the scattering length to the box length is of order $1/N$, i.e., the parameter Na/L is kept fixed. Changing a with N can be done by scaling, i.e., we write

$$v(|\mathbf{x}|) = \frac{1}{a^2}v_1(|\mathbf{x}|/a) \tag{5.3}$$

for some v_1 having scattering length 1, and vary a while keeping v_1 fixed.[1] It is easily checked that the v so defined has scattering length a. It is important to note that, in the limit considered, a tends to zero (as $N^{-2/3}$ since $L = (N/\rho)^{1/3} \sim N^{1/3}$ for ρ fixed), and v becomes a *hard* potential of *short* range. This is the *opposite* of the usual mean field limit where the strength of the potential goes to zero while its range tends to infinity.

We shall refer to this as the *Gross-Pitaevskii (GP) limit* since Na/L will turn out to be the natural interaction parameter for inhomogeneous Bose gases confined in traps, that are described by the Gross-Pitaevskii equation discussed in Chapters 6 and 7. Its significance for a homogeneous gas can also be seen by noting that Na/L is the ratio of ρa to $1/L^2$, i.e., in the GP limit the interaction energy per particle is of the same order of magnitude as the energy gap in the box, so that the interaction is still clearly visible, even though $a \to 0$. Note that $\rho a^3 \sim N^{-2}$ in the GP limit, so letting $N \to \infty$ with ρ fixed and Na/L fixed can be regarded as a *simultaneous thermodynamic and low density limit*. For simplicity, we shall here treat only the 3D case.

Theorem 5.1 (BEC in a dilute limit). *Assume that, as $N \to \infty$, $\rho = N/L^3$ and $g = Na/L$ stay fixed, and impose either periodic or Neumann boundary conditions for H_N. Then*

$$\lim_{N\to\infty} \frac{1}{N}\frac{1}{L^3} \iint \gamma(\mathbf{x}, \mathbf{y})d\mathbf{x}d\mathbf{y} = 1 . \tag{5.4}$$

The reason we do not deal with Dirichlet boundary conditions at this point should be clear from the discussion preceding the theorem: There would be an additional contribution $\sim 1/L^2$ to the energy, of the same order as the interaction

---

[1] By scaling, this is mathematically equivalent to fixing the interaction potential v (and therefore fixing a) but taking $L \sim N$, i.e., $\rho = N/L^3 \sim N^{-2}$.

energy, and the system would not be homogeneous any more. Dirichlet boundary conditions can, however, be treated with the methods of Chapter 7.

By scaling, the limit in Theorem 5.1 is equivalent to considering a Bose gas in a *fixed box* of side length $L = 1$, and keeping Na fixed as $N \to \infty$, i.e., $a \sim 1/N$. The ground state energy of the system is then, asymptotically, $N \times 4\pi Na$, and Theorem 5.1 implies that the one-particle reduced density matrix γ of the ground state converges, after division by N, to the projection onto the constant function. An analogous result holds true for inhomogeneous systems as will be discussed in Chapter 7.

The proof of Theorem 5.1 has two main ingredients. One is *localization of the energy* that is stated as Lemma 5.2 below. This lemma is a refinement of the energy estimates of Section 2.2 and says essentially that the kinetic energy of the ground state is concentrated in a subset of configuration space where at least one pair of particles is close together and whose volume tends to zero as $a \to 0$. The other is the generalized Poincaré inequality, Lemma 4.1, from which one deduces that the one-particle density matrix is approximately constant if the kinetic energy is localized in a small set.

The localization lemma will be proved in a slightly more general version than is necessary for Theorem 5.1, namely with the gradient ∇ replaced by $\nabla_\varphi = \nabla + i(0, 0, \varphi/L)$, cf. Eq. (4.3). We denote by H'_N the corresponding many-body Hamiltonian (2.1) with ∇_φ in place of ∇. This generalization will be used in the subsequent discussion of superfluidity, but a reader who wishes to focus on Theorem 5.1 only can simply ignore the φ and the reference to the diamagnetic inequality in the proof. We denote the gradient with respect to $\mathbf{x}_1$ by ∇_1, and the corresponding modified operator by $\nabla_{1,\varphi}$.

Lemma 5.2 (Localization of energy). *Let $\mathcal{K}$ be a box of side length L. For all symmetric, normalized wave functions $\Psi(\mathbf{x}_1, \ldots, \mathbf{x}_N)$ with periodic boundary conditions on $\mathcal{K}$, and for $N \geq Y^{-1/17}$,*

$$\frac{1}{N}\langle \Psi, H'_N \Psi \rangle \geq \left(1 - \text{const.} \, Y^{1/17}\right)$$
$$\times \left(4\pi\mu\rho a + \mu \int_{\mathcal{K}^{N-1}} d\mathbf{X} \int_{\Omega_\mathbf{X}} d\mathbf{x}_1 |\nabla_{1,\varphi}\Psi(\mathbf{x}_1, \mathbf{X})|^2\right),$$

$$(5.5)$$

where $\mathbf{X} = (\mathbf{x}_2, \ldots, \mathbf{x}_N)$, $d\mathbf{X} = \prod_{j=2}^{N} d\mathbf{x}_j$, and

$$\Omega_\mathbf{X} = \left\{\mathbf{x}_1 : \min_{j \geq 2} |\mathbf{x}_1 - \mathbf{x}_j| \geq R\right\} \qquad (5.6)$$

with $R = aY^{-5/17}$.

Proof. Since Ψ is symmetric, the left side of (5.5) can be written as

$$\int_{\mathcal{K}^{N-1}} d\mathbf{X} \int_{\mathcal{K}} d\mathbf{x}_1 \left[\mu |\nabla_{1,\varphi} \Psi(\mathbf{x}_1, \mathbf{X})|^2 + \tfrac{1}{2} \sum_{j \geq 2} v(|\mathbf{x}_1 - \mathbf{x}_j|) |\Psi(\mathbf{x}_1, \mathbf{X})|^2 \right] . \qquad (5.7)$$

For any $\varepsilon > 0$ and $R > 0$ this is

$$\geq \varepsilon T + (1 - \varepsilon)(T^{\mathrm{in}} + I) + (1 - \varepsilon)T_\varphi^{\mathrm{out}} , \qquad (5.8)$$

with

$$T = \mu \int_{\mathcal{K}^{N-1}} d\mathbf{X} \int_{\mathcal{K}} d\mathbf{x}_1 \big|\nabla_1 |\Psi(\mathbf{x}_1, \mathbf{X})|\big|^2 , \qquad (5.9)$$

$$T^{\mathrm{in}} = \mu \int_{\mathcal{K}^{N-1}} d\mathbf{X} \int_{\Omega_{\mathbf{X}}^c} d\mathbf{x}_1 \big|\nabla_1 |\Psi(\mathbf{x}_1, \mathbf{X})|\big|^2 , \qquad (5.10)$$

$$T_\varphi^{\mathrm{out}} = \mu \int_{\mathcal{K}^{N-1}} d\mathbf{X} \int_{\Omega_{\mathbf{X}}} d\mathbf{x}_1 \big|\nabla_{1,\varphi} \Psi(\mathbf{x}_1, \mathbf{X})\big|^2 , \qquad (5.11)$$

and

$$I = \tfrac{1}{2} \int_{\mathcal{K}^{N-1}} d\mathbf{X} \int_{\mathcal{K}} d\mathbf{x}_1 \sum_{j \geq 2} v(|\mathbf{x}_1 - \mathbf{x}_j|) |\Psi(\mathbf{x}_1, \mathbf{X})|^2 . \qquad (5.12)$$

Here

$$\Omega_{\mathbf{X}}^c = \{\mathbf{x}_1 : |\mathbf{x}_1 - \mathbf{x}_j| < R \text{ for some } j \geq 2\} \qquad (5.13)$$

is the complement of $\Omega_{\mathbf{X}}$, and the diamagnetic inequality [LLo] $|\nabla_\varphi f(\mathbf{x})|^2 \geq |\nabla |f(\mathbf{x})||^2$ has been used. The proof is completed by using the estimates used for the proof of Theorem 2.4, in particular (2.65) and (2.67)–(2.70), which tell us that for $\varepsilon = Y^{1/17}$ and $R = aY^{-5/17}$

$$\varepsilon T + (1 - \varepsilon)(T^{\mathrm{in}} + I) \geq \left(1 - \text{const.} \, Y^{1/17}\right) 4\pi\mu\rho a \qquad (5.14)$$

as long as $N \geq Y^{-1/17}$. □

Proof of Theorem 5.1. We combine Lemma 5.2 (with $\varphi = 0$ and hence $H'_N = H_N$) with Lemma 4.1 that gives a lower bound to the second term on the right side of (5.5). We thus infer that, for any symmetric Ψ with $\langle \Psi, \Psi \rangle = 1$ and for N large enough,

$$\frac{1}{N} \langle \Psi, H_N \Psi \rangle \left(1 - \text{const.} \, Y^{1/17}\right)^{-1}$$

$$\geq 4\pi\mu\rho a - CY^{1/17} \left(\frac{1}{L^2} - \frac{1}{N} \langle \Psi, \sum_j \nabla_j^2 \Psi \rangle \right)$$

$$+ \frac{c}{L^2} \int_{\mathcal{K}^{N-1}} d\mathbf{X} \int_{\mathcal{K}} d\mathbf{x}_1 \Big| \Psi(\mathbf{x}_1, \mathbf{X}) - L^{-3} \big[\int_{\mathcal{K}} d\mathbf{x} \Psi(\mathbf{x}, \mathbf{X}) \big] \Big|^2 ,$$

$$(5.15)$$

where we used that $|\Omega^c| \leq \frac{4\pi}{3} NR^3 = \text{const.} \, L^3 Y^{2/17}$. For the ground state wave function Ψ_0 the energy per particle, $N^{-1}\langle \Psi_0, H_N \Psi_0 \rangle$, is bounded from above by $4\pi\mu\rho a(1 + \text{const} \, Y^{1/3})$ according to (2.14). The same holds for the kinetic energy per particle, $-N^{-1}\langle \Psi_0, \sum_j \nabla_j^2 \Psi_0 \rangle$ that appears on the right side of (5.15). We now multiply the inequality (5.15) by L^2 and use the upper bound (2.14). The terms $4\pi\mu\rho a L^2$ on both sides of the resulting inequality cancel. For the remaining terms we note that in the limit considered $\rho a L^2 = \text{const.}$ while $Y \to 0$. Hence the positive last term has to vanish in the limit and because $c > 0$ this means

$$\lim_{N \to \infty} \int_{\mathcal{K}^{N-1}} d\mathbf{X} \int_{\mathcal{K}} d\mathbf{x}_1 \Big| \Psi_0(\mathbf{x}_1, \mathbf{X}) - L^{-3} \big[\int_{\mathcal{K}} d\mathbf{x} \Psi_0(\mathbf{x}, \mathbf{X}) \big] \Big|^2 = 0 \,. \tag{5.16}$$

This proves (5.4), since

$$\int_{\mathcal{K}^{N-1}} d\mathbf{X} \int_{\mathcal{K}} d\mathbf{x}_1 \Big| \Psi_0(\mathbf{x}_1, \mathbf{X}) - L^{-3} \big[\int_{\mathcal{K}} d\mathbf{x} \Psi_0(\mathbf{x}, \mathbf{X}) \big] \Big|^2$$
$$= 1 - \frac{1}{NL^3} \int_{\mathcal{K} \times \mathcal{K}} \gamma(\mathbf{x}, \mathbf{x}') d\mathbf{x} d\mathbf{x}' \,. \tag{5.17}$$

$\square$

5.2 Superfluidity

The phenomenological two-fluid model of superfluidity (see, e.g., [TT]) is based on the idea that the particle density ρ is composed of two parts, the density ρ_s of the inviscid superfluid and the normal fluid density ρ_n. If an external velocity field is imposed on the fluid (for instance by moving the walls of the container) only the viscous normal component responds to the velocity field, while the superfluid component stays at rest. In accord with these ideas the superfluid density in the ground state is often defined as follows [HoM]: Let E_0 denote the ground state energy of the system in the rest frame and E_0' the ground state energy, measured in the moving frame, when a velocity field $\mathbf{v}$ is imposed. Then for small $\mathbf{v}$

$$\frac{E_0'}{N} = \frac{E_0}{N} + (\rho_s/\rho)\frac{1}{2}m\mathbf{v}^2 + O(|\mathbf{v}|^4) \tag{5.18}$$

where N is the particle number and m the particle mass. At positive temperatures the ground state energy should be replaced by the free energy. (Remark: It is important here that (5.18) holds uniformly for all large N; i.e., that the error term $O(|\mathbf{v}|^4)$ can be bounded independently of N. For fixed N and a finite box, Eq. (5.18) with $\rho_s/\rho = 1$ always holds for a Bose gas with an arbitrary interaction if $\mathbf{v}$ is small enough, owing to the discreteness of the energy spectrum.[2]) There are other definitions of the superfluid density that may lead to different results [PrSv], but this is the one we shall use here. We shall not dwell on this issue since it is not clear that there is a "one-size-fits-all" definition of superfluidity. For instance, in the definition we use here the ideal Bose gas is a perfect superfluid in its ground state, whereas the definition of Landau in terms of a linear dispersion relation of elementary excitations would indicate otherwise. Our main result is that with the definition adopted here there is 100% superfluidity in the ground state of a 3D Bose gas in the GP limit explained in the previous section.

One of the unresolved issues in the theory of superfluidity is its relation to Bose-Einstein condensation (BEC). It has been argued that in general neither condition is necessary for the other (c.f., e.g., [H, ABCG, KT]), but in the case considered here, i.e., the GP limit of a 3D gas, we show that 100% BEC into the constant wave function (in the rest frame) prevails even if an external velocity field is imposed. A simple example illustrating the fact that BEC is not necessary for superfluidity is the 1D hard-core Bose gas. This system is well known to have a spectrum like that of an ideal Fermi gas [Gi2] (see also Chapter 8 and Appendix B), and it is easy to see that it is superfluid in its ground state in the sense of (5.18). On the other hand, it has no BEC [Le, PiSt]. The definition of the superfluid velocity as the gradient of the phase of the condensate wave function [HoM, Bm] is clearly not applicable in such cases.

We consider a Bose gas with the Hamiltonian (2.1) in a box $\mathcal{K}$ of side length L, assuming periodic boundary conditions in all three coordinate directions. Imposing an external velocity field $\mathbf{v} = (0, 0, \pm|\mathbf{v}|)$ means that the momentum operator $\mathbf{p} = -i\hbar\nabla$ is replaced by $\mathbf{p} - m\mathbf{v}$, retaining the periodic boundary conditions. The Hamiltonian in the moving frame is thus

$$H'_N = -\mu \sum_{j=1}^{N} \nabla^2_{j,\varphi} + \sum_{1 \leq i < j \leq N} v(|\mathbf{x}_i - \mathbf{x}_j|) , \qquad (5.19)$$

where $\nabla_{j,\varphi} = \nabla_j + i(0, 0, \varphi/L)$ and the dimensionless phase φ is connected to the velocity $\mathbf{v}$ by

$$\varphi = \frac{\pm|\mathbf{v}|Lm}{\hbar} . \qquad (5.20)$$

---

[2]The ground state with $\mathbf{v} = 0$ remains an eigenstate of the Hamiltonian with arbitrary $\mathbf{v}$ (but not necessarily a ground state) since its total momentum is zero. Its energy is $\frac{1}{2}mN\mathbf{v}^2$ above the ground state energy for $\mathbf{v} = 0$. Since in a finite box the spectrum of the Hamiltonian for arbitrary $\mathbf{v}$ is discrete and the energy gap above the ground state is bounded away from zero for $\mathbf{v}$ small, the ground state for $\mathbf{v} = 0$ is at the same time the ground state of the Hamiltonian with $\mathbf{v}$ if $\frac{1}{2}mN\mathbf{v}^2$ is smaller than the gap.

Let $E_0(N, a, \varphi)$ denote the ground state energy of (5.19) with periodic boundary conditions. Obviously it is no restriction to consider only the case $-\pi \leq \varphi \leq \pi$, since E_0 is periodic in φ with period 2π (see Remark 1 below). For Ψ_0 the ground state of H'_N, let γ_N be its one-particle reduced density matrix . We are interested in the *Gross-Pitaevskii* (GP) limit $N \to \infty$ with Na/L fixed. We also fix the box size L. This means that a should vary like $1/N$ which, as explained in the previous section, can be achieved by writing $v(r) = a^{-2}v_1(r/a)$, where v_1 is a fixed potential with scattering length 1, while a changes with N.

Theorem 5.3 (Superfluidity and BEC of homogeneous gas). *For $|\varphi| \leq \pi$*

$$\lim_{N\to\infty} \frac{E_0(N, a, \varphi)}{N} = 4\pi\mu a\rho + \mu\frac{\varphi^2}{L^2} \tag{5.21}$$

in the limit $N \to \infty$ with Na/L and L fixed. Here $\rho = N/L^3$, so $a\rho$ is fixed too. In the same limit, for $|\varphi| < \pi$,

$$\lim_{N\to\infty} \frac{1}{N} \gamma_N(\mathbf{x}, \mathbf{x}') = \frac{1}{L^3} \tag{5.22}$$

in trace class norm, i.e., $\lim_{N\to\infty} \mathrm{tr}\left[\left|\gamma_N/N - |L^{-3/2}\rangle\langle L^{-3/2}|\right|\right] = 0$.

Note that, by the definition (5.18) of ρ_s and Eq. (5.20), Eq. (5.21) means that $\rho_s = \rho$, i.e., there is 100% superfluidity. For $\varphi = 0$, Eq. (5.21) follows from Eq. (2.8), while (5.22) for $\varphi = 0$ is the BEC of Theorem 5.1.[3]

Remarks: 1. By a unitary gauge transformation,

$$\left(U\Psi\right)(\mathbf{x}_1, \ldots, \mathbf{x}_N) = e^{\mathrm{i}\varphi(\sum_i z_i)/L} \Psi(\mathbf{x}_1, \ldots, \mathbf{x}_N), \tag{5.23}$$

the passage from (2.1) to (5.19) is equivalent to replacing periodic boundary conditions in a box by the *twisted boundary condition*

$$\Psi(\mathbf{x}_1 + (0, 0, L), \mathbf{x}_2, \ldots, \mathbf{x}_N) = e^{\mathrm{i}\varphi}\Psi(\mathbf{x}_1, \mathbf{x}_2, \ldots, \mathbf{x}_N) \tag{5.24}$$

in the direction of the velocity field, while retaining the original Hamiltonian (2.1).

2. The criterion $|\varphi| \leq \pi$ means that $|\mathbf{v}| \leq \pi\hbar/(mL)$. The corresponding energy $\frac{1}{2}m(\pi\hbar/(mL))^2$ is the gap in the excitation spectrum of the one-particle Hamiltonian in the finite-size system.

---

[3]The convention in Theorem 5.1, where ρ and Na/L stay fixed, is different from the one employed here, where L and Na/L are fixed, but these two conventions are clearly equivalent by scaling.

3. The reason that we have to restrict ourselves to $|\varphi| < \pi$ in the second part of Theorem 5.3 is that for $|\varphi| = \pi$ there are two ground states of the operator $(\nabla + i\varphi/L)^2$ with periodic boundary conditions. All we can say in this case is that there is a subsequence of γ_N that converges to a density matrix of rank ≤ 2, whose range is spanned by these two functions

Proof of Theorem 5.3. As in the proof of Theorem 5.1 we combine the localization Lemma 5.2, this time with $\varphi \neq 0$, and a generalized Poincaré inequality , this time Lemma 4.2. We thus infer that, for any symmetric Ψ with $\langle \Psi, \Psi \rangle = 1$ and for N large enough,

$$\frac{1}{N} \langle \Psi, H'_N \Psi \rangle \left(1 - \text{const.} \, Y^{1/17} \right)^{-1}$$

$$\geq 4\pi\mu\rho a + \mu \frac{\varphi^2}{L^2} - C Y^{1/17} \left(\frac{1}{L^2} - \frac{1}{N} \langle \Psi, \sum_j \nabla^2_{j,\varphi} \Psi \rangle \right)$$

$$+ \frac{c}{L^2} \int_{\mathcal{K}^{N-1}} d\mathbf{X} \int_{\mathcal{K}} d\mathbf{x}_1 \Big| \Psi(\mathbf{x}_1, \mathbf{X}) - L^{-3} [\int_{\mathcal{K}} d\mathbf{x} \Psi(\mathbf{x}, \mathbf{X})] \Big|^2 ,$$

where we used that $|\Omega^c| \leq \frac{4\pi}{3} N R^3 = \text{const.} \, L^3 Y^{2/17}$. From this we can infer two things. First, since the kinetic energy, divided by N, is certainly bounded independently of N, as the upper bound shows, we get that

$$\liminf_{N \to \infty} \frac{E_0(N, a, \varphi)}{N} \geq 4\pi\mu\rho a + \mu \frac{\varphi^2}{L^2} \tag{5.25}$$

for any $|\varphi| < \pi$. By continuity this holds also for $|\varphi| = \pi$, proving (5.21). (To be precise, $E_0/N - \mu\varphi^2 L^{-2}$ is concave in φ, and therefore stays concave, and in particular continuous, in the limit $N \to \infty$.) Secondly, since the upper and the lower bounds to E_0 agree in the limit considered, the positive last term in (5.15) has to vanish in the limit. I.e., we get that for the ground state wave function Ψ_0 of H'_N

$$\lim_{N \to \infty} \int_{\mathcal{K}^{N-1}} d\mathbf{X} \int_{\mathcal{K}} d\mathbf{x}_1 \Big| \Psi_0(\mathbf{x}_1, \mathbf{X}) - L^{-3} [\int_{\mathcal{K}} d\mathbf{x} \Psi_0(\mathbf{x}, \mathbf{X})] \Big|^2 = 0 . \tag{5.26}$$

Using again (5.17), this proves (5.22) in a weak sense. As explained in [LSe, LSSY], this suffices for the convergence $N^{-1}\gamma_N \to |L^{-3/2}\rangle\langle L^{-3/2}|$ in trace class norm. $\square$

Theorem 5.3 can be generalized in various ways to a physically more realistic setting, for example replacing the periodic box by a cylinder centered at the origin. We shall comment on such extensions at the end of Chapter 7.

Chapter 6

Gross-Pitaevskii Equation for Trapped Bosons

In the recent experiments on Bose condensation (see, e.g., [KD]), the particles are confined at very low temperatures in a 'trap' where the particle density is *inhomogeneous*, contrary to the case of a large 'box', where the density is essentially uniform. We model the trap by a slowly varying confining potential V, with $V(\mathbf{x}) \to \infty$ as $|\mathbf{x}| \to \infty$. The Hamiltonian becomes

$$H = \sum_{i=1}^{N} \{-\mu \Delta_i + V(\mathbf{x}_i)\} + \sum_{1 \leq i < j \leq N} v(|\mathbf{x}_i - \mathbf{x}_j|) . \tag{6.1}$$

Shifting the energy scale if necessary we can assume that V is nonnegative. The ground state energy, $\hbar \omega$, of $-\mu \Delta + V(\mathbf{x})$ is a natural energy unit and the corresponding length unit, $\sqrt{\hbar/(m\omega)} = \sqrt{2\mu/(\hbar\omega)} \equiv L_{\mathrm{osc}}$, is a measure of the extension of the trap.

   In the sequel we shall be considering a limit where a/L_{osc} tends to zero while $N \to \infty$. Experimentally a/L_{osc} can be changed in two ways: One can either vary L_{osc} or a. The first alternative is usually simpler in practice but very recently a direct tuning of the scattering length itself has also been shown to be feasible [CCRCW]. Mathematically, both alternatives are equivalent, of course. The first corresponds to writing $V(\mathbf{x}) = L_{\mathrm{osc}}^{-2} V_1(\mathbf{x}/L_{\mathrm{osc}})$ and keeping V_1 and v fixed. The second corresponds to writing the interaction potential as $v(|\mathbf{x}|) = a^{-2} v_1(|\mathbf{x}|/a)$ like in (5.3), where v_1 has unit scattering length, and keeping V and v_1 fixed. This is equivalent to the first, since for given V_1 and v_1 the ground state energy of (6.1), measured in units of $\hbar\omega$, depends only on N and a/L_{osc}. In the dilute limit when a

is much smaller than the mean particle distance, the energy becomes independent of v_1.

We choose L_{osc} as a length unit. The energy unit is $\hbar\omega = 2\mu L_{\text{osc}}^{-2} = 2\mu$. Moreover, we find it convenient to regard V and v_1 as fixed. This justifies the notion $E_0(N, a)$ for the quantum mechanical ground state energy.

The idea is now to use the information about the thermodynamic limiting energy of the dilute Bose gas in a box to find the ground state energy of (6.1) in an appropriate limit. This has been done in [LSeY1, LSeY2] and in this chapter we give an account of this work. As we saw in Chapters 2 and 3 there is a difference in the ρ dependence between two and three dimensions, so we can expect a related difference now. We discuss 3D first.

6.1   Three Dimensions

Associated with the quantum mechanical ground state energy problem is the Gross-Pitaevskii (GP) energy functional [Gr1, Gr2, Pi]

$$\mathcal{E}^{\text{GP}}[\phi] = \int_{\mathbb{R}^3} \left(\mu|\nabla\phi|^2 + V|\phi|^2 + 4\pi\mu a|\phi|^4 \right) d\mathbf{x} \qquad (6.2)$$

with the subsidiary condition

$$\int_{\mathbb{R}^3} |\phi|^2 = N. \qquad (6.3)$$

As before, $a > 0$ is the scattering length of v. The corresponding energy is

$$E^{\text{GP}}(N, a) = \inf_{\int |\phi|^2 = N} \mathcal{E}^{\text{GP}}[\phi] = \mathcal{E}^{\text{GP}}[\phi^{\text{GP}}], \qquad (6.4)$$

with a unique, positive ϕ^{GP}. The existence of the minimizer ϕ^{GP} is proved by standard techniques and it can be shown to be continuously differentiable, see [LSeY1], Sect. 2 and Appendix A. The minimizer depends on N and a, of course, and when this is important we denote it by $\phi^{\text{GP}}_{N,a}$.

The variational equation satisfied by the minimizer is the *GP equation*

$$-\mu\Delta\phi^{\text{GP}}(\mathbf{x}) + V(\mathbf{x})\phi^{\text{GP}}(\mathbf{x}) + 8\pi\mu a\phi^{\text{GP}}(\mathbf{x})^3 = \mu^{\text{GP}}\phi^{\text{GP}}(\mathbf{x}), \qquad (6.5)$$

where μ^{GP} is the chemical potential, given by

$$\mu^{\text{GP}} = dE^{\text{GP}}(N, a)/dN = E^{\text{GP}}(N, a)/N + (4\pi\mu a/N) \int |\phi^{\text{GP}}(\mathbf{x})|^4 d\mathbf{x}. \qquad (6.6)$$

The GP theory has the following scaling property:

$$E^{\text{GP}}(N, a) = NE^{\text{GP}}(1, Na), \qquad (6.7)$$

and

$$\phi^{GP}_{N,a}(\mathbf{x}) = N^{1/2}\phi^{GP}_{1,Na}(\mathbf{x}).\tag{6.8}$$

Hence we see that the relevant parameter in GP theory is the combination Na.

We now turn to the relation of E^{GP} and ϕ^{GP} to the quantum mechanical ground state. If $v = 0$, then the ground state of (6.1) is

$$\Psi_0(\mathbf{x}_1,\ldots,\mathbf{x}_N) = \prod_{i=1}^{N}\phi_0(\mathbf{x}_i)$$

with ϕ_0 the normalized ground state of $-\mu\Delta + V(\mathbf{x})$. In this case clearly $\phi^{GP} = \sqrt{N}\,\phi_0$, and then $E^{GP} = N\hbar\omega = E_0$. In the other extreme, if $V(\mathbf{x}) = 0$ for $\mathbf{x}$ inside a large box of volume L^3 and $V(\mathbf{x}) = \infty$ otherwise, then $\phi^{GP} \approx \sqrt{N/L^3}$ and we get $E^{GP}(N,a) = 4\pi\mu aN^2/L^3$, which is the previously considered energy E_0 for the homogeneous gas in the low density regime. (In this case, the gradient term in $\mathcal{E}^{GP}$ plays no role.)

In general, we expect that for *dilute* gases in a suitable limit

$$E_0 \approx E^{GP} \quad \text{and} \quad \rho^{QM}(\mathbf{x}) \approx \left|\phi^{GP}(\mathbf{x})\right|^2 \equiv \rho^{GP}(\mathbf{x}),\tag{6.9}$$

where the quantum mechanical particle density in the ground state is defined by

$$\rho^{QM}(\mathbf{x}) = N\int |\Psi_0(\mathbf{x},\mathbf{x}_2,\ldots,\mathbf{x}_N)|^2 d\mathbf{x}_2\cdots d\mathbf{x}_N.\tag{6.10}$$

Dilute means here that

$$\bar{\rho}a^3 \ll 1,\tag{6.11}$$

where

$$\bar{\rho} = \frac{1}{N}\int |\rho^{GP}(\mathbf{x})|^2 d\mathbf{x}\tag{6.12}$$

is the *mean density*.

The limit in which (6.9) can be expected to be true should be chosen so that *all three* terms in $\mathcal{E}^{GP}$ make a contribution. The scaling relations (6.7) and (6.8) indicate that fixing Na as $N \to \infty$ is the right thing to do (and this is quite relevant since experimentally N can be quite large, 10^6 and more, and Na can range from about 1 to 10^4 [DGPS]). Fixing Na (which we refer to as the GP case) also means that we really are dealing with a dilute limit, because the mean density $\bar{\rho}$ is then of the order N (since $\bar{\rho}_{N,a} = N\bar{\rho}_{1,Na}$) and hence

$$a^3\bar{\rho} \sim N^{-2}.\tag{6.13}$$

The precise statement of (6.9) is:

Theorem 6.1 (GP limit of the QM ground state energy and density). *If $N \to \infty$ with Na fixed, then*

$$\lim_{N \to \infty} \frac{E_0(N,a)}{E^{GP}(N,a)} = 1, \tag{6.14}$$

and

$$\lim_{N \to \infty} \frac{1}{N} \rho_{N,a}^{QM}(\mathbf{x}) = |\phi_{1,Na}^{GP}(\mathbf{x})|^2 \tag{6.15}$$

in the weak L^1-sense.

Convergence can not only be proved for the ground state energy and density, but also for the individual energy components:

Theorem 6.2 (Asymptotics of the energy components). *Let ψ_0 denote the solution to the zero-energy scattering equation for v (under the boundary condition $\lim_{|\mathbf{x}| \to \infty} \psi_0(\mathbf{x}) = 1$) and $s = \int |\nabla \psi_0|^2/(4\pi a)$. Then $0 < s \le 1$ and, in the same limit as in Theorem 6.1 above,*

$$\lim_{N \to \infty} \int |\nabla_{\mathbf{x}_1} \Psi_0(\mathbf{x}_1, \mathbf{X})|^2 d\mathbf{x}_1 \, d\mathbf{X}$$
$$= \int |\nabla \phi_{1,Na}^{GP}(\mathbf{x})|^2 d\mathbf{x} + 4\pi Nas \int |\phi_{1,Na}^{GP}(\mathbf{x})|^4 d\mathbf{x}, \tag{6.16a}$$

$$\lim_{N \to \infty} \int V(\mathbf{x}_1)|\Psi_0(\mathbf{x}_1, \mathbf{X})|^2 d\mathbf{x}_1 \, d\mathbf{X} = \int V(\mathbf{x})|\phi_{1,Na}^{GP}(\mathbf{x})|^2 d\mathbf{x}, \tag{6.16b}$$

$$\lim_{N \to \infty} \frac{1}{2} \sum_{j=2}^{N} \int v(|\mathbf{x}_1 - \mathbf{x}_j|)|\Psi_0(\mathbf{x}_1, \mathbf{X})|^2 d\mathbf{x}_1 \, d\mathbf{X}$$
$$= (1 - s)4\pi Na \int |\phi_{1,Na}^{GP}(\mathbf{x})|^4 d\mathbf{x}. \tag{6.16c}$$

Here we introduced again the short hand notation (1.18). Theorem 6.2 is a simple consequence of Theorem 6.1 by variation with respect to the different components of the energy, as was also noted in [CS2]. More precisely, Eq. (6.14) can be written as

$$\lim_{N \to \infty} \frac{1}{N} E_0(N,a) = E^{GP}(1, Na). \tag{6.17}$$

The ground state energy is a concave function of the mass parameter μ, so it is legitimate to differentiate both sides of (6.17) with respect to μ. In doing so, it has to be noted that Na depends on μ through the scattering length. Using (2.13) one sees that

$$\frac{d(\mu a)}{d\mu} = \frac{1}{4\pi} \int |\nabla \psi_0|^2 d\mathbf{x} \tag{6.18}$$

by the Feynman-Hellmann principle, since ψ_0 minimizes the left side of (2.13).

We remark that in the case of a two-dimensional Bose gas, where the relevant parameter to be kept fixed in the GP limit is $N/|\ln(a^2\bar{\rho}_N)|$ (c.f. Chapter 3 and Section 6.2), the parameter s in Theorem 6.2 can be shown to be always equal to 1. I.e., in 2D the interaction energy is purely kinetic in the GP limit (see [CS1]).

To describe situations where Na is very large, it is appropriate to consider a limit where, as $N \to \infty$, $a \gg N^{-1}$, i.e. $Na \to \infty$, but still $\bar{\rho}a^3 \to 0$. In this case, the gradient term in the GP functional becomes negligible compared to the other terms and the so-called *Thomas-Fermi (TF) functional*

$$\mathcal{E}^{\mathrm{TF}}[\rho] = \int_{\mathbb{R}^3} \left(V\rho + 4\pi\mu a\rho^2 \right) d\mathbf{x} \tag{6.19}$$

arises. (Note that this functional has nothing to do with the fermionic theory invented by Thomas and Fermi in 1927, except for a certain formal analogy.) It is defined for nonnegative functions ρ on $\mathbb{R}^3$. Its ground state energy E^{TF} and density ρ^{TF} are defined analogously to the GP case. (The TF functional is especially relevant for the two-dimensional Bose gas. There a has to decrease exponentially with N in the GP limit, so the TF limit is more adequate; see Section 6.2 below).

Our second main result of this chapter is that minimization of (6.19) reproduces correctly the ground state energy and density of the many-body Hamiltonian in the limit when $N \to \infty$, $a^3\bar{\rho} \to 0$, but $Na \to \infty$ (which we refer to as the TF case), provided the external potential is reasonably well behaved. We will assume that V is asymptotically equal to some function W that is homogeneous of some order $s > 0$, i.e., $W(\lambda\mathbf{x}) = \lambda^s W(\mathbf{x})$ for all $\lambda > 0$, and locally Hölder continuous (see [LSeY2] for a precise definition). This condition can be relaxed, but it seems adequate for most practical applications and simplifies things considerably.

Theorem 6.3 (TF limit of the QM ground state energy and density). *Assume that V satisfies the conditions stated above. If $g \equiv Na \to \infty$ as $N \to \infty$, but still $a^3\bar{\rho} \to 0$, then*

$$\lim_{N\to\infty} \frac{E_0(N,a)}{E^{\mathrm{TF}}(N,a)} = 1, \tag{6.20}$$

and

$$\lim_{N\to\infty} \frac{g^{3/(s+3)}}{N} \rho_{N,a}^{\mathrm{QM}}(g^{1/(s+3)}\mathbf{x}) = \tilde{\rho}_{1,1}^{\mathrm{TF}}(\mathbf{x}) \tag{6.21}$$

in the weak L^1-sense, where $\tilde{\rho}_{1,1}^{\mathrm{TF}}$ is the minimizer of the TF functional under the condition $\int \rho = 1$, $a = 1$, and with V replaced by W.

In the following, we will present the essentials of the proofs Theorems 6.1 and 6.3. We will derive appropriate upper and lower bounds on the ground state energy E_0.

The proof of the lower bound in Theorem 6.1 presented here is a modified version of (and partly simpler than) the original proof in [LSeY1].

The convergence of the densities follows from the convergence of the energies in the usual way by variation with respect to the external potential. For simplicity, we set $\mu \equiv 1$ in the following.

Proof of Theorems 6.1 and 6.3. Part 1: Upper bound to the QM energy. To derive an upper bound on E_0 we use a generalization of a trial wave function of Dyson [D1], who used this function to give an upper bound on the ground state energy of the homogeneous hard core Bose gas (c.f. Section 2.1). It is of the form

$$\Psi(\mathbf{x}_1, \ldots, \mathbf{x}_N) = \prod_{i=1}^{N} \phi^{\mathrm{GP}}(\mathbf{x}_i) F(\mathbf{x}_1, \ldots, \mathbf{x}_N), \qquad (6.22)$$

where F is constructed in the following way:

$$F(\mathbf{x}_1, \ldots, \mathbf{x}_N) = \prod_{i=1}^{N} f(t_i(\mathbf{x}_1, \ldots, \mathbf{x}_i)), \qquad (6.23)$$

where $t_i = \min\{|\mathbf{x}_i - \mathbf{x}_j|, 1 \le j \le i-1\}$ is the distance of $\mathbf{x}_i$ to its *nearest neighbor* among the points $\mathbf{x}_1, \ldots, \mathbf{x}_{i-1}$, and f is a function of $r \ge 0$. As in (2.19) we choose it to be

$$f(r) = \begin{cases} f_0(r)/f_0(b) & \text{for} \quad r < b \\ 1 & \text{for} \quad r \ge b, \end{cases} \qquad (6.24)$$

where f_0 is the solution of the zero energy scattering equation (2.3) and b is some cut-off parameter of order $b \sim \bar{\rho}^{-1/3}$. The function (6.22) is not totally symmetric, but for an upper bound it is nevertheless an acceptable test wave function since the bosonic ground state energy is equal to the *absolute* ground state energy.

The result of a somewhat lengthy computation, similar to the one given in Section 2.1 (see [LSeY1] for details), is the upper bound

$$E_0(N, a) \le E^{\mathrm{GP}}(N, a) \left(1 + O(a\bar{\rho}^{1/3})\right). \qquad (6.25)$$

Part 2: Lower bound to the QM energy, GP case. To obtain a lower bound for the QM ground state energy the strategy is to divide space into boxes and use the estimate on the homogeneous gas, given in Theorem 2.4, in each box with

Neumann boundary conditions. One then minimizes over all possible divisions of the particles among the different boxes. This gives a lower bound to the energy because discontinuous wave functions for the quadratic form defined by the Hamiltonian are now allowed. We can neglect interactions among particles in different boxes because $v \geq 0$. Finally, one lets the box size tend to zero. However, it is not possible to simply approximate V by a constant potential in each box. To see this consider the case of noninteracting particles, i.e., $v = 0$ and hence $a = 0$. Here $E_0 = N\hbar\omega$, but a 'naive' box method gives only $\min_{\mathbf{x}} V(\mathbf{x})$ as lower bound, since it clearly pays to put all the particles with a constant wave function in the box with the lowest value of V.

For this reason we start by separating out the GP wave function in each variable and write a general wave function Ψ as

$$\Psi(\mathbf{x}_1, \ldots, \mathbf{x}_N) = \prod_{i=1}^{N} \phi^{\mathrm{GP}}(\mathbf{x}_i) F(\mathbf{x}_1, \ldots, \mathbf{x}_N). \tag{6.26}$$

Here $\phi^{\mathrm{GP}} = \phi_{N,a}^{\mathrm{GP}}$ is normalized so that $\int |\phi^{\mathrm{GP}}|^2 = N$. Eq. (6.26) defines F for a given Ψ because ϕ^{GP} is everywhere strictly positive, being the ground state of the operator $-\Delta + V + 8\pi a |\phi^{\mathrm{GP}}|^2$. We now compute the expectation value of H in the state Ψ. Using partial integration and the variational equation (6.5) for ϕ^{GP}, we see that

$$\frac{\langle \Psi | H \Psi \rangle}{\langle \Psi | \Psi \rangle} - E^{\mathrm{GP}}(N, a) = 4\pi a \int |\rho^{\mathrm{GP}}|^2 + Q(F), \tag{6.27}$$

with

$$Q(F) = \sum_{i=1}^{N} \frac{\int \prod_{k=1}^{N} \rho^{\mathrm{GP}}(\mathbf{x}_k) \left(|\nabla_i F|^2 + \left[\frac{1}{2} \sum_{j \neq i} v(|\mathbf{x}_i - \mathbf{x}_j|) - 8\pi a \rho^{\mathrm{GP}}(\mathbf{x}_i) \right] |F|^2 \right)}{\int \prod_{k=1}^{N} \rho^{\mathrm{GP}}(\mathbf{x}_k) |F|^2}. \tag{6.28}$$

We recall that $\rho^{\mathrm{GP}}(\mathbf{x}) = |\phi_{N,a}^{\mathrm{GP}}(\mathbf{x})|^2$. For computing the ground state energy of H we have to minimize the normalized quadratic form Q. Compared to the expression for the energy involving Ψ itself we have thus obtained the replacements

$$V(\mathbf{x}) \to -8\pi a \rho^{\mathrm{GP}}(\mathbf{x}) \quad \text{and} \quad \prod_{i=1}^{N} d\mathbf{x}_i \to \prod_{i=1}^{N} \rho^{\mathrm{GP}}(\mathbf{x}_i) d\mathbf{x}_i. \tag{6.29}$$

We now use the box method on *this* problem. More precisely, labeling the boxes by an index α, we have

$$\inf_F Q(F) \geq \inf_{\{n_\alpha\}} \sum_\alpha \inf_{F_\alpha} Q_\alpha(F_\alpha), \tag{6.30}$$

where Q_α is defined by the same formula as Q but with the integrations limited to the box α, F_α is a wave function with particle number n_α, and the infimum is taken over all distributions of the particles with $\sum n_\alpha = N$.

We now fix some $M > 0$, that will eventually tend to ∞, and restrict ourselves to boxes inside a cube Λ_M of side length M. Since $v \geq 0$ the contribution to (6.30) of boxes outside this cube is easily estimated from below by $-8\pi N a \sup_{\mathbf{x} \notin \Lambda_M} \rho^{\mathrm{GP}}(\mathbf{x})$, which, divided by N, is arbitrarily small for M large, since Na is fixed and $\phi^{\mathrm{GP}}/N^{1/2} = \phi^{\mathrm{GP}}_{1,Na}$ decreases faster than exponentially at infinity ([LSeY1], Lemma A.5).

For the boxes inside the cube Λ_M we want to use Lemma 2.5 and therefore we must approximate ρ^{GP} by constants in each box. Let $\rho_{\alpha,\max}$ and $\rho_{\alpha,\min}$, respectively, denote the maximal and minimal values of ρ^{GP} in box α. Define

$$\Psi_\alpha(\mathbf{x}_1, \ldots, \mathbf{x}_{n_\alpha}) = F_\alpha(\mathbf{x}_1, \ldots, \mathbf{x}_{n_\alpha}) \prod_{k=1}^{n_\alpha} \phi^{\mathrm{GP}}(\mathbf{x}_k), \tag{6.31}$$

and

$$\Psi_\alpha^{(i)}(\mathbf{x}_1, \ldots, \mathbf{x}_{n_\alpha}) = F_\alpha(\mathbf{x}_1, \ldots, \mathbf{x}_{n_\alpha}) \prod_{\substack{k=1 \\ k \neq i}}^{n_\alpha} \phi^{\mathrm{GP}}(\mathbf{x}_k). \tag{6.32}$$

We have, for all $1 \leq i \leq n_\alpha$,

$$\int \prod_{k=1}^{n_\alpha} \rho^{\mathrm{GP}}(\mathbf{x}_k) \left(|\nabla_i F_\alpha|^2 + \tfrac{1}{2} \sum_{j \neq i} v(|\mathbf{x}_i - \mathbf{x}_j|) |F_\alpha|^2 \right)$$
$$\geq \rho_{\alpha,\min} \int \left(|\nabla_i \Psi_\alpha^{(i)}|^2 + \tfrac{1}{2} \sum_{j \neq i} v(|\mathbf{x}_i - \mathbf{x}_j|) |\Psi_\alpha^{(i)}|^2 \right). \tag{6.33}$$

We now use Lemma 2.5 to get, for all $0 \leq \varepsilon \leq 1$,

$$(6.33) \geq \rho_{\alpha,\min} \int \left(\varepsilon |\nabla_i \Psi_\alpha^{(i)}|^2 + a(1 - \varepsilon) U(t_i) |\Psi_\alpha^{(i)}|^2 \right) \tag{6.34}$$

where t_i is the distance to the nearest neighbor of $\mathbf{x}_i$, c.f., (2.42), and U the potential (2.43).

Since $\Psi_\alpha = \phi^{\mathrm{GP}}(\mathbf{x}_i) \Psi_\alpha^{(i)}$ we can estimate

$$|\nabla_i \Psi_\alpha|^2 \leq 2\rho_{\alpha,\max} |\nabla_i \Psi_\alpha^{(i)}|^2 + 2|\Psi_\alpha^{(i)}|^2 N C_M \tag{6.35}$$

with

$$CM = \frac{1}{N} \sup_{\mathbf{x} \in \Lambda_M} |\nabla \phi^{\mathrm{GP}}(\mathbf{x})|^2 = \sup_{\mathbf{x} \in \Lambda_M} |\nabla \phi^{\mathrm{GP}}_{1,Na}(\mathbf{x})|^2. \qquad (6.36)$$

Since Na is fixed, C_M is independent of N. Inserting (6.35) into (6.34), summing over i and using $\rho^{\mathrm{GP}}(\mathbf{x}_i) \le \rho_{\alpha,\max}$ in the last term of (6.28) (in the box α), we get

$$Q_\alpha(F_\alpha) \ge \frac{\rho_{\alpha,\min}}{\rho_{\alpha,\max}} E_\varepsilon^U(n_\alpha, L) - 8\pi a \rho_{\alpha,\max} n_\alpha - \varepsilon C_M n_\alpha, \qquad (6.37)$$

where L is the side length of the box and $E_\varepsilon^U(n_\alpha, L)$ is the ground state energy of

$$\sum_{i=1}^{n_\alpha} (-\tfrac{1}{2}\varepsilon \Delta_i + (1 - \varepsilon)aU(t_i)) \qquad (6.38)$$

in the box (c.f. (2.48)). We want to minimize (6.37) with respect to n_α and drop the subsidiary condition $\sum_\alpha n_\alpha = N$ in (6.30). This can only lower the minimum. For the time being we also ignore the last term in (6.37). (The total contribution of this term for all boxes is bounded by $\varepsilon C_M N$ and will be shown to be negligible compared to the other terms.)

Since the lower bound for the energy of Theorem 2.4 was obtained precisely from a lower bound to the operator (6.38), we can use the statement and proof of Theorem 2.4. From this we see that

$$E_\varepsilon^U(n_\alpha, L) \ge (1 - \varepsilon)\frac{4\pi a n_\alpha^2}{L^3}(1 - CY_\alpha^{1/17}) \qquad (6.39)$$

with $Y_\alpha = a^3 n_\alpha/L^3$, provided Y_α is small enough, and that $\varepsilon \ge Y_\alpha^{1/17}$ and $n_\alpha \ge (\text{const.})Y_\alpha^{-1/17}$. The condition on ε is certainly fulfilled if we choose $\varepsilon = Y^{1/17}$ with $Y = a^3 N/L^3$. We now want to show that the n_α minimizing the right side of (6.37) is large enough for (6.39) to apply.

If the minimum of the right side of (6.37) (without the last term) is taken for some $\bar{n}_\alpha$, we have

$$\frac{\rho_{\alpha,\min}}{\rho_{\alpha,\max}}\left(E_\varepsilon^U(\bar{n}_\alpha + 1, L) - E_\varepsilon^U(\bar{n}_\alpha, L)\right) \ge 8\pi a \rho_{\alpha,\max}. \qquad (6.40)$$

On the other hand, we claim that

Lemma 6.4. *For any n*

$$E_\varepsilon^U(n + 1, L) - E_\varepsilon^U(n, L) \le 8\pi a \frac{n}{L^3}. \qquad (6.41)$$

Proof. Denote the operator (6.38) by $\tilde{H}_n$, with $n_\alpha = n$, and let $\tilde{\Psi}_n$ be its ground state. Let t_i' be the distance to the nearest neighbor of $\mathbf{x}_i$ among the $n+1$ points $\mathbf{x}_1, \ldots, \mathbf{x}_{n+1}$ (without $\mathbf{x}_i$) and t_i the corresponding distance excluding $\mathbf{x}_{n+1}$. Obviously, for $1 \leq i \leq n$,

$$U(t_i') \leq U(t_i) + U(|\mathbf{x}_i - \mathbf{x}_{n+1}|) \tag{6.42}$$

and

$$U(t_{n+1}') \leq \sum_{i=1}^{n} U(|\mathbf{x}_i - \mathbf{x}_{n+1}|). \tag{6.43}$$

Therefore

$$\tilde{H}_{n+1} \leq \tilde{H}_n - \tfrac{1}{2}\varepsilon\Delta_{n+1} + 2a\sum_{i=1}^{n} U(|\mathbf{x}_i - \mathbf{x}_{n+1}|). \tag{6.44}$$

Using $\tilde{\Psi}_n/L^{3/2}$ as trial function for $\tilde{H}_{n+1}$ we arrive at (6.41).   □

Eq. (6.41) together with (6.40) shows that $\bar{n}_\alpha$ is at least $\sim \rho_{\alpha,\max}L^3$. We shall choose $L \sim N^{-1/10}$, so the conditions needed for (6.39) are fulfilled for N large enough, since $\rho_{\alpha,\max} \sim N$ and hence $\bar{n}_\alpha \sim N^{7/10}$ and $Y_\alpha \sim N^{-2}$.

In order to obtain a lower bound on Q_α we therefore have to minimize

$$4\pi a \left(\frac{\rho_{\alpha,\min}}{\rho_{\alpha,\max}} \frac{n_\alpha^2}{L^3} \left(1 - CY^{1/17}\right) - 2n_\alpha\rho_{\alpha,\max} \right). \tag{6.45}$$

We can drop the requirement that n_α has to be an integer. The minimum of (6.45) is obtained for

$$n_\alpha = \frac{\rho_{\alpha,\max}^2}{\rho_{\alpha,\min}} \frac{L^3}{(1 - CY^{1/17})}. \tag{6.46}$$

By Eq. (6.27) this gives the following lower bound, including now the last term in (6.37) as well as the contributions from the boxes outside Λ_M,

$$E_0(N, a) - E^{\mathrm{GP}}(N, a) \geq$$

$$4\pi a \int |\rho^{\mathrm{GP}}|^2 - 4\pi a \sum_{\alpha \subset \Lambda_M} \rho_{\alpha,\min}^2 L^3 \left(\frac{\rho_{\alpha,\max}^3}{\rho_{\alpha,\min}^3} \frac{1}{(1 - CY^{1/17})} \right) \tag{6.47}$$

$$- Y^{1/17} N C_M - 4\pi a N \sup_{\mathbf{x} \notin \Lambda_M} \rho^{\mathrm{GP}}(\mathbf{x}).$$

Now ρ^{GP} is differentiable and strictly positive. Since all the boxes are in the fixed cube Λ_M there are constants $C' < \infty$, $C'' > 0$, such that

$$\rho_{\alpha,\max} - \rho_{\alpha,\min} \leq NC'L, \qquad \rho_{\alpha,\min} \geq NC''. \tag{6.48}$$

Since $L \sim N^{-1/10}$ and $Y \sim N^{-17/10}$ we therefore have, for large N,

$$\frac{\rho_{\alpha,\max}^3}{\rho_{\alpha,\min}^3} \frac{1}{(1 - CY^{1/17})} \leq 1 + (\text{const.})N^{-1/10} \tag{6.49}$$

Also,

$$4\pi a \sum_{\alpha \subset \Lambda_M} \rho_{\alpha,\min}^2 L^3 \leq 4\pi a \int |\rho^{\text{GP}}|^2 \leq E^{\text{GP}}(N, a). \tag{6.50}$$

Hence, noting that $E^{\text{GP}}(N, a) = NE^{\text{GP}}(1, Na) \sim N$ since Na is fixed,

$$\frac{E_0(N, a)}{E^{\text{GP}}(N, a)} \geq 1 - (\text{const.})(1 + C_M)N^{-1/10} - (\text{const.}) \sup_{\mathbf{x} \notin \Lambda_M} |\phi_{1,Na}^{\text{GP}}|^2, \tag{6.51}$$

where the constants depend on Na. We can now take $N \to \infty$ and then $M \to \infty$.

Part 3: Lower bound to the QM energy, TF case. In the above proof of the lower bound in the GP case we did not attempt to keep track of the dependence of the constants on Na. In the TF case $Na \to \infty$, so one would need to take a closer look at this dependence if one wanted to carry the proof directly over to this case. But we don't have to do so, because there is a simpler direct proof. Using the explicit form of the TF minimizer, namely

$$\rho_{N,a}^{\text{TF}}(\mathbf{x}) = \frac{1}{8\pi a}[\mu^{\text{TF}} - V(\mathbf{x})]_+, \tag{6.52}$$

where $[t]_+ \equiv \max\{t, 0\}$ and μ^{TF} is chosen so that the normalization condition $\int \rho_{N,a}^{\text{TF}} = N$ holds, we can use

$$V(\mathbf{x}) \geq \mu^{\text{TF}} - 8\pi a \rho^{\text{TF}}(\mathbf{x}) \tag{6.53}$$

to get a replacement as in (6.29), but without changing the measure. Moreover, ρ^{TF} has compact support, so, applying again the box method described above, the boxes far out do not contribute to the energy. However, μ^{TF} (which depends only on the combination Na) tends to infinity as $Na \to \infty$. We need to control the asymptotic behavior of μ^{TF}, and this leads to the restrictions on V described in the paragraph preceding Theorem 6.3. For simplicity, we shall here only consider the case when V itself is homogeneous, i.e., $V(\lambda \mathbf{x}) = \lambda^s V(\mathbf{x})$ for all $\lambda > 0$ with some $s > 0$.

In the same way as in (6.6) we have, with $g = Na$,

$$\mu^{\text{TF}}(g) = dE^{\text{TF}}(N, a)/dN = E^{\text{TF}}(1, g) + 4\pi g \int |\rho_{1,g}^{\text{TF}}(\mathbf{x})|^2 d\mathbf{x}. \tag{6.54}$$

The TF energy, chemical potential and minimizer satisfy the scaling relations

$$E^{\mathrm{TF}}(1, g) = g^{s/(s+3)} E^{\mathrm{TF}}(1, 1), \tag{6.55}$$

$$\mu^{\mathrm{TF}}(g) = g^{s/(s+3)} \mu^{\mathrm{TF}}(1), \tag{6.56}$$

and

$$g^{3/(s+3)} \rho_{1,g}^{\mathrm{TF}}(g^{1/(s+3)}\mathbf{x}) = \rho_{1,g}^{\mathrm{TF}}(\mathbf{x}). \tag{6.57}$$

We also introduce the scaled interaction potential, $\widehat{v}$, by

$$\widehat{v}(\mathbf{x}) = g^{2/(s+3)} v(g^{1/(s+3)}\mathbf{x}) \tag{6.58}$$

with scattering length

$$\widehat{a} = g^{-1/(s+3)} a. \tag{6.59}$$

Using (6.53), (6.54) and the scaling relations we obtain

$$E_0(N, a) \geq E^{\mathrm{TF}}(N, a) + 4\pi N g^{s/(s+3)} \int |\rho_{1,1}^{\mathrm{TF}}|^2 + g^{-2/(s+3)} Q \tag{6.60}$$

with

$$Q = \inf_{\int |\Psi|^2 = 1} \sum_i \int \left(|\nabla_i \Psi|^2 + \tfrac{1}{2} \sum_{j \neq i} \widehat{v}(\mathbf{x}_i - \mathbf{x}_j) |\Psi|^2 - 8\pi N \widehat{a} \rho_{1,1}^{\mathrm{TF}}(\mathbf{x}_i) |\Psi|^2 \right). \tag{6.61}$$

We can now proceed exactly as in Part 2 to arrive at the analogy of Eq. (6.47), which in the present case becomes

$$E_0(N, a) - E^{\mathrm{TF}}(N, a) \geq$$
$$4\pi N g^{s/(s+3)} \int |\rho_{1,1}^{\mathrm{TF}}|^2 - 4\pi N \widehat{a} \sum_\alpha \rho_{\alpha,\max}^2 L^3 (1 - C\widehat{Y}^{1/17})^{-1}. \tag{6.62}$$

Here $\rho_{\alpha,\max}$ is the maximum of $\rho_{1,1}^{\mathrm{TF}}$ in the box α, and $\widehat{Y} = \widehat{a}^3 N / L^3$. This holds as long as L does not decrease too fast with N. In particular, if L is simply fixed, this holds for all large enough N. Note that

$$\bar{\rho} = N \bar{\rho}_{1,g} \sim N g^{-3/(s+3)} \bar{\rho}_{1,1}, \tag{6.63}$$

so that $\widehat{a}^3 N \sim a^3 \bar{\rho}$ goes to zero as $N \to \infty$ by assumption. Hence, if we first let $N \to \infty$ (which implies $\widehat{Y} \to 0$) and then take L to zero, we arrive at the desired result

$$\liminf_{N \to \infty} \frac{E_0(N, a)}{E^{\mathrm{TF}}(N, a)} \geq 1 \tag{6.64}$$

in the limit $N \to \infty$, $a^3 \bar{\rho} \to 0$. Here we used the fact that (because V, and hence ρ^{TF}, is continuous by assumption) the Riemann sum $\sum_\alpha \rho_{\alpha,\max}^2 L^3$ converges to $\int |\rho_{1,1}^{\mathrm{TF}}|^2$ as $L \to 0$. Together with the upper bound (6.25) and the fact that $E^{\mathrm{GP}}(N,a)/E^{\mathrm{TF}}(N,a) = E^{\mathrm{GP}}(1,Na)/E^{\mathrm{TF}}(1,Na) \to 1$ as $Na \to \infty$, which holds under our regularity assumption on V (c.f. Lemma 2.3 in [LSeY2]), this proves (6.14) and (6.20).

Part 4: Convergence of the densities. The convergence of the energies implies the convergence of the densities in the usual way by variation of the external potential. We consider the TF case here; the GP case is analogous. Set again $g = Na$. Making the replacement

$$V(\mathbf{x}) \longrightarrow V(\mathbf{x}) + \delta g^{s/(s+3)} Z(g^{-1/(s+3)} \mathbf{x}) \tag{6.65}$$

for some positive $Z \in C_0^\infty$ and redoing the upper and lower bounds we see that (6.20) holds with W replaced by $W + \delta Z$. Differentiating with respect to δ at $\delta = 0$ yields

$$\lim_{N \to \infty} \frac{g^{3/(s+3)}}{N} \rho_{N,a}^{\mathrm{QM}}(g^{1/(s+3)} \mathbf{x}) = \tilde{\rho}_{1,1}^{\mathrm{TF}}(\mathbf{x}) \tag{6.66}$$

in the sense of distributions. Since the functions all have L^1-norm 1, we can conclude that there is even weak L^1-convergence. $\qquad\square$

6.2 Two Dimensions

In contrast to the three-dimensional case the energy per particle for a dilute gas in two dimensions is *nonlinear* in ρ. In view of Schick's formula (3.1) for the energy of the homogeneous gas it would appear natural to take the interaction into account in two dimensional GP theory by a term

$$4\pi \int_{\mathbb{R}^2} |\ln(|\phi(\mathbf{x})|^2 a^2)|^{-1} |\phi(\mathbf{x})|^4 d\mathbf{x}, \tag{6.67}$$

and such a term has, indeed, been suggested in [Sh] and [KNSQ]. However, since the nonlinearity appears only in a logarithm, this term is unnecessarily complicated as far as leading order computations are concerned. For dilute gases it turns out to be sufficient, to leading order, to use an interaction term of the same form as in the three-dimensional case, i.e, define the GP functional as (for simplicity we put $\mu = 1$ in this section)

$$\mathcal{E}^{\mathrm{GP}}[\phi] = \int_{\mathbb{R}^2} \left(|\nabla\phi|^2 + V|\phi|^2 + 4\pi\alpha|\phi|^4 \right) d\mathbf{x}, \tag{6.68}$$

where, instead of a, the coupling constant is now

$$\alpha = |\ln(\bar{\rho}_N a^2)|^{-1} \tag{6.69}$$

with $\bar{\rho}_N$ the *mean density* for the GP functional at coupling constant 1 and particle number N. This is defined analogously to (6.12) as

$$\bar{\rho}_N = \frac{1}{N} \int |\phi_{N,1}^{\text{GP}}|^4 d\mathbf{x} \tag{6.70}$$

where $\phi_{N,1}^{\text{GP}}$ is the minimizer of (6.68) with $\alpha = 1$ and subsidiary condition $\int |\phi|^2 = N$. Note that α in (6.69) depends on N through the mean density.

Let us denote the GP energy for a given N and coupling constant α by $E^{\text{GP}}(N, \alpha)$ and the corresponding minimizer by $\phi_{N,\alpha}^{\text{GP}}$. As in three dimensions the scaling relations

$$E^{\text{GP}}(N, \alpha) = N E^{\text{GP}}(1, N\alpha) \tag{6.71}$$

and

$$N^{-1/2} \phi_{N,\alpha}^{\text{GP}} = \phi_{1,N\alpha}^{\text{GP}} \tag{6.72}$$

hold, and the relevant parameter is

$$g \equiv N\alpha. \tag{6.73}$$

In three dimensions, where $\alpha = a$, it is natural to consider the limit $N \to \infty$ with $g = Na =$ const. The analogue of Theorem 6.1 in two dimensions is

Theorem 6.5 (Two-dimensional GP limit theorem). *If, for $N \to \infty$, $a^2 \bar{\rho}_N \to 0$ with $g = N/|\ln(a^2 \bar{\rho}_N)|$ fixed, then*

$$\lim_{N \to \infty} \frac{E_0(N, a)}{E^{\text{GP}}(N, 1/|\ln(a^2 \bar{\rho}_N)|)} = 1 \tag{6.74}$$

and

$$\lim_{N \to \infty} \frac{1}{N} \rho_{N,a}^{\text{QM}}(\mathbf{x}) = |\phi_{1,g}^{\text{GP}}(\mathbf{x})|^2 \tag{6.75}$$

in the weak L^1-sense.

This result, however, is of rather limited use in practice. The reason is that in two dimensions the scattering length has to decrease exponentially with N if g is fixed. The parameter g is typically *very large* in two dimensions so it is more appropriate to consider the limit $N \to \infty$ and $g \to \infty$ (but still $\bar{\rho}_N a^2 \to 0$).

For potentials V that are *homogeneous* functions of $\mathbf{x}$, i.e.,

$$V(\lambda \mathbf{x}) = \lambda^s V(\mathbf{x}) \tag{6.76}$$

for some $s > 0$, this limit can be described by the a 'Thomas-Fermi' energy functional like (6.19) with coupling constant unity:

$$\mathcal{E}^{\mathrm{TF}}[\rho] = \int_{\mathbb{R}^2} \left(V(\mathbf{x})\rho(\mathbf{x}) + 4\pi\rho(\mathbf{x})^2\right) d\mathbf{x}. \tag{6.77}$$

This is just the GP functional without the gradient term and $\alpha = 1$. Here ρ is a nonnegative function on $\mathbb{R}^2$ and the normalization condition is

$$\int \rho(\mathbf{x})d\mathbf{x} = 1. \tag{6.78}$$

The minimizer of (6.77) can be given explicitly. It is

$$\rho_{1,1}^{\mathrm{TF}}(\mathbf{x}) = (8\pi)^{-1}[\mu^{\mathrm{TF}} - V(\mathbf{x})]_+ \tag{6.79}$$

where the chemical potential μ^{TF} is determined by the normalization condition (6.78) and $[t]_+ = t$ for $t \geq 0$ and zero otherwise. We denote the corresponding energy by $E^{\mathrm{TF}}(1,1)$. By scaling one obtains

$$\lim_{g\to\infty} E^{\mathrm{GP}}(1,g)/g^{s/(s+2)} = E^{\mathrm{TF}}(1,1), \tag{6.80}$$

$$\lim_{g\to\infty} g^{2/(s+2)}\rho_{1,g}^{\mathrm{GP}}(g^{1/(s+2)}\mathbf{x}) = \rho_{1,1}^{\mathrm{TF}}(\mathbf{x}), \tag{6.81}$$

with the latter limit in the strong L^2 sense.

Our main result about two-dimensional Bose gases in external potentials satisfying (6.76) is that analogous limits also hold for the many-particle quantum mechanical ground state at low densities:

Theorem 6.6 (Two-dimensional TF limit theorem). *In 2D, if $a^2\bar\rho_N \to 0$, but $g = N/|\ln(\bar\rho_N a^2)| \to \infty$ as $N \to \infty$, then*

$$\lim_{N\to\infty} \frac{E_0(N,a)}{g^{s/s+2}} = E^{\mathrm{TF}}(1,1) \tag{6.82}$$

and, in the weak L^1 sense,

$$\lim_{N\to\infty} \frac{g^{2/(s+2)}}{N}\rho_{N,a}^{\mathrm{QM}}(g^{1/(s+2)}\mathbf{x}) = \rho_{1,1}^{\mathrm{TF}}(\mathbf{x}). \tag{6.83}$$

Remarks: 1. As in Theorem 6.3, it is sufficient that V is asymptotically equal to some homogeneous potential, W. In this case, $E^{\mathrm{TF}}(1,1)$ and $\rho_{1,1}^{\mathrm{TF}}$ in Theorem 6.6 should be replaced by the corresponding quantities for W.

2. From Eq. (6.81) it follows that

$$\bar\rho_N \sim N^{s/(s+2)} \tag{6.84}$$

for large N. Hence the low density criterion $a^2\bar{\rho} \ll 1$, means that $a/L_{\rm osc} \ll N^{-s/2(s+2)}$.

We shall now comment briefly on the proofs of Theorems 6.5 and 6.6, mainly pointing out the differences from the 3D case considered previously.

The upper bounds for the energy are obtained exactly in a same way as in three dimensions. For the lower bound in Theorem 6.5 the point to notice is that the expression (6.45), that has to be minimized over n_α, is in 2D replaced by

$$4\pi \left(\frac{\rho_{\alpha,\min}}{\rho_{\alpha,\max}} \frac{n_\alpha^2}{L^2} \frac{1}{|\ln(a^2 n_\alpha/L^2)|} \left(1 - \frac{C}{|\ln(a^2 N/L^2)|^{1/5}} \right) - \frac{2n_\alpha \rho_{\alpha,\max}}{|\ln(a^2 \bar{\rho}_N)|} \right), \quad (6.85)$$

since Eq. (6.39) has to be replaced by the analogous inequality for 2D (c.f. (3.31)). To minimize (6.85) we use the following lemma:

Lemma 6.7. *For $0 < x, b < 1$ and $k \geq 1$ we have*

$$\frac{x^2}{|\ln x|} - 2\frac{b}{|\ln b|} xk \geq -\frac{b^2}{|\ln b|} \left(1 + \frac{1}{(2|\ln b|)^2} \right) k^2. \quad (6.86)$$

Proof. Replacing x by xk and using the monotonicity of $\ln$ we see that it suffices to consider $k = 1$. Since $\ln x \geq -\frac{1}{de} x^{-d}$ for all $d > 0$ we have

$$\frac{x^2}{b^2} \frac{|\ln b|}{|\ln x|} - 2\frac{x}{b} \geq \frac{|\ln b|}{b^2} e dx^{2+d} - \frac{2x}{b} \geq c(d)(b^d e d |\ln b|)^{-1/(1+d)} \quad (6.87)$$

with

$$c(d) = 2^{(2+d)/(1+d)} \left(\frac{1}{(2+d)^{(2+d)/(1+d)}} - \frac{1}{(2+d)^{1/(1+d)}} \right) \geq -1 - \frac{1}{4} d^2. \quad (6.88)$$

Choosing $d = 1/|\ln b|$ gives the desired result.   □

Applying this lemma with $x = a^2 n_\alpha/L^2$, $b = a^2 \rho_{\alpha,\max}$ and

$$k = \frac{\rho_{\alpha,\max}}{\rho_{\alpha,\min}} \left(1 - \frac{C}{|\ln(a^2 N/L^2)|^{1/5}} \right)^{-1} \frac{|\ln(a^2 \rho_{\alpha,\max})|}{|\ln(a^2 \bar{\rho}_N)|} \quad (6.89)$$

we get the bound

$$(6.85) \geq -4\pi \frac{\rho_{\alpha,\max}^2 L^2}{|\ln(a^2 \bar{\rho}_N)|} \left(1 + \frac{1}{4|\ln(a^2 \rho_{\alpha,\max})|^2} \right) k. \quad (6.90)$$

In the limit considered, k and the factor in parenthesis both tend to 1 and the Riemann sum over the boxes α converges to the integral as $L \to 0$.

The TF case, Thm. 6.6, is treated in the same way as in three dimensions, with modifications analogous to those just discussed when passing from 3D to 2D in GP theory.

Chapter 7

Bose-Einstein Condensation and Superfluidity for Dilute Trapped Gases

It was shown in the previous chapter that, for each fixed Na, the minimization of the GP functional correctly reproduces the large N asymptotics of the ground state energy and density of H – but no assertion about BEC in this limit was made. We will now extend this result by showing that in the Gross-Pitaevskii limit there is indeed 100% Bose condensation in the ground state. This is a generalization of the homogeneous case considered in Theorem 5.1 and although it is not the same as BEC in the thermodynamic limit it is quite relevant for the actual experiments with Bose gases in traps. In the following, we concentrate on the 3D case, but analogous considerations apply also to the 2D case. We also discuss briefly some extensions of Theorem 5.3 pertaining to superfluidity in trapped gases.

As in the last chapter we choose to keep the length scale L_{osc} of the confining potential fixed and thus write Na instead of Na/L_{osc}. Consequently the powers of N appearing in the proofs are different from those in the proof Theorem 5.1, where we kept Na/L and N/L^3 fixed.

For later use, we define the projector

$$P^{\mathrm{GP}} = |\phi^{\mathrm{GP}}\rangle\langle\phi^{\mathrm{GP}}| . \tag{7.1}$$

Here (and everywhere else in this chapter) we denote $\phi^{\mathrm{GP}} \equiv \phi_{1,Na}^{\mathrm{GP}}$ for simplicity, where $\phi_{1,Na}^{\mathrm{GP}}$ is the minimizer of the GP functional (6.2) with parameter Na and normalization condition $\int |\phi|^2 = 1$ (compare with (6.8)). Moreover, we set $\mu \equiv 1$.

In the following, Ψ_0 denotes the (nonnegative and normalized) ground state of the Hamiltonian (6.1). BEC refers to the reduced one-particle density matrix $\gamma(\mathbf{x}, \mathbf{x}')$ of Ψ_0, defined in (5.1). The precise definition of BEC is that for some $c > 0$ this integral operator has for all large N an eigenfunction with eigenvalue $\geq cN$.

Complete (or 100%) BEC is defined to be the property that $\frac{1}{N}\gamma(\mathbf{x}, \mathbf{x}')$ not only has an eigenvalue of order one, as in the general case of an incomplete BEC, but in the limit it has only one nonzero eigenvalue (namely 1). Thus, $\frac{1}{N}\gamma(\mathbf{x}, \mathbf{x}')$ becomes a simple product $\varphi(\mathbf{x})\varphi(\mathbf{x}')^*$ as $N \to \infty$, in which case φ is called the *condensate wave function*. In the GP limit, i.e., $N \to \infty$ with Na fixed, we can show that this is the case, and the condensate wave function is, in fact, the GP minimizer ϕ^{GP}.

Theorem 7.1 (Bose-Einstein condensation in a trap). *For each fixed Na*

$$\lim_{N \to \infty} \frac{1}{N}\gamma(\mathbf{x}, \mathbf{x}') = \phi^{\mathrm{GP}}(\mathbf{x})\phi^{\mathrm{GP}}(\mathbf{x}') \ .$$

in trace norm, i.e., $\mathrm{tr}\left|\frac{1}{N}\gamma - P^{\mathrm{GP}}\right| \to 0$.

We remark that Theorem 7.1 implies that there is also 100% condensation for all n-particle reduced density matrices

$$\gamma^{(n)}(\mathbf{x}_1, \ldots, \mathbf{x}_n; \mathbf{x}_1', \ldots, \mathbf{x}_n')$$
$$= n!\binom{N}{n}\int \Psi_0(\mathbf{x}_1, \ldots, \mathbf{x}_N)\Psi_0(\mathbf{x}_1', \ldots, \mathbf{x}_n', \mathbf{x}_{n+1}, \ldots \mathbf{x}_N)d\mathbf{x}_{n+1}\cdots d\mathbf{x}_N$$

$$(7.2)$$

of Ψ_0, i.e., they converge, after division by the normalization factor, to the one-dimensional projector onto the n-fold tensor product of ϕ^{GP}. In other words, for n fixed particles the probability of finding them all in the same state ϕ^{GP} tends to 1 in the limit considered. To see this, let a^*, a denote the boson creation and annihilation operators for the state ϕ^{GP}, and observe that

$$1 \geq \lim_{N \to \infty} N^{-n}\langle\Psi_0|(a^*)^n a^n|\Psi_0\rangle = \lim_{N \to \infty} N^{-n}\langle\Psi_0|(a^*a)^n|\Psi_0\rangle \ , \qquad (7.3)$$

since the terms coming from the commutators $[a, a^*] = 1$ are of lower order as $N \to \infty$ and vanish in the limit. From convexity it follows that

$$N^{-n}\langle\Psi_0|(a^*a)^n|\Psi_0\rangle \geq N^{-n}\langle\Psi_0|a^*a|\Psi_0\rangle^n \qquad (7.4)$$

which converges to 1 as $N \to \infty$, proving our claim.

Another corollary, important for the interpretation of experiments, concerns the momentum distribution of the ground state.

Corollary 7.2 (Convergence of momentum distribution). *Let*

$$\widehat{\rho}(\mathbf{k}) = \int \int \gamma(\mathbf{x}, \mathbf{x}') \exp[i\mathbf{k} \cdot (\mathbf{x} - \mathbf{x}')]d\mathbf{x}d\mathbf{x}'$$

denote the one-particle momentum density of Ψ_0. *Then, for fixed* Na,

$$\lim_{N \to \infty} \frac{1}{N}\widehat{\rho}(\mathbf{k}) = |\widehat{\phi}^{\mathrm{GP}}(\mathbf{k})|^2$$

strongly in $L^1(\mathbb{R}^3)$. *Here,* $\widehat{\phi}^{\mathrm{GP}}$ *denotes the Fourier transform of* ϕ^{GP}.

Proof. If $\mathcal{F}$ denotes the (unitary) operator 'Fourier transform' and if h is an arbitrary L^∞-function, then

$$\left| \frac{1}{N} \int \widehat{\rho}h - \int |\widehat{\phi}^{\mathrm{GP}}|^2 h \right| = \left| \mathrm{tr}[\mathcal{F}^{-1}(\gamma/N - P^{\mathrm{GP}})\mathcal{F}h] \right|$$

$$\leq \|h\|_\infty \, \mathrm{tr} \, |\gamma/N - P^{\mathrm{GP}}|,$$

from which we conclude that

$$\|\widehat{\rho}/N - |\widehat{\phi}^{\mathrm{GP}}|^2\|_1 \leq \mathrm{tr} \, |\gamma/N - P^{\mathrm{GP}}| \, . \qquad \square$$

As already stated, Theorem 7.1 is a generalization of Theorem 5.1, the latter corresponding to the case that V is a box potential. It should be noted, however, that we use different scaling conventions in these two theorems: In Theorem 5.1 the box size grows as $N^{1/3}$ to keep the density fixed, while in Theorem 7.1 we choose to keep the confining external potential fixed. Both conventions are equivalent, of course, c.f. the remarks in the second paragraph of Chapter 6, but when comparing the exponents of N that appear in the proofs of the two theorems the different conventions should be born in mind.

As in Theorem 5.1 there are two essential components of our proof of Theorem 7.1. The first is a proof that the part of the kinetic energy that is associated with the interaction v (namely, the second term in (6.16a)) is mostly located in small balls surrounding each particle. More precisely, these balls can be taken to have radius roughly $N^{-5/9}$, which is much smaller than the mean-particle spacing $N^{-1/3}$. (The exponents differ from those of Lemma 5.2 because of different scaling conventions.) This allows us to conclude that the function of $\mathbf{x}$ defined for each fixed value of $\mathbf{X}$ by

$$f_{\mathbf{X}}(\mathbf{x}) = \frac{1}{\phi^{\mathrm{GP}}(\mathbf{x})} \Psi_0(\mathbf{x}, \mathbf{X}) \geq 0 \tag{7.5}$$

has the property that $\nabla_{\mathbf{x}} f_{\mathbf{X}}(\mathbf{x})$ is almost zero outside the small balls centered at points of $\mathbf{X}$.

The complement of the small balls has a large volume but it can be a weird set; it need not even be connected. Therefore, the smallness of $\nabla_{\mathbf{x}} f_{\mathbf{X}}(\mathbf{x})$ in this set does not guarantee that $f_{\mathbf{X}}(\mathbf{x})$ is nearly constant (in $\mathbf{x}$), or even that it is continuous. We need $f_{\mathbf{X}}(\mathbf{x})$ to be nearly constant in order to conclude BEC. What saves the day is the knowledge that the total kinetic energy of $f_{\mathbf{X}}(\mathbf{x})$ (including the balls) is not huge. The result that allows us to combine these two pieces of information in order to deduce the almost constancy of $f_{\mathbf{X}}(\mathbf{x})$ is the generalized Poincaré inequality in Lemma 4.3. The important point in this lemma is that there is no restriction on Ω concerning regularity or connectivity.

Using the results of Theorem 6.2, partial integration and the GP equation (i.e., the variational equation for ϕ^{GP}, see Eq. (6.5)) we see that

$$\lim_{N\to\infty} \int |\phi^{\mathrm{GP}}(\mathbf{x})|^2 |\nabla_{\mathbf{x}} f_{\mathbf{X}}(\mathbf{x})|^2 d\mathbf{x}\, d\mathbf{X} = 4\pi N a s \int |\phi^{\mathrm{GP}}(\mathbf{x})|^4 d\mathbf{x} \ . \tag{7.6}$$

The following Lemma shows that to leading order all the energy in (7.6) is concentrated in small balls.

Lemma 7.3 (Localization of the energy in a trap). *For fixed* $\mathbf{X}$ *let*

$$\Omega_{\mathbf{X}} = \left\{ \mathbf{x} \in \mathbb{R}^3 \ \middle| \ \min_{k\geq 2} |\mathbf{x} - \mathbf{x}_k| \geq N^{-1/3-\delta} \right\} \tag{7.7}$$

for some $0 < \delta < 2/9$. *Then*

$$\lim_{N\to\infty} \int d\mathbf{X} \int_{\Omega_{\mathbf{X}}} d\mathbf{x} |\phi^{\mathrm{GP}}(\mathbf{x})|^2 |\nabla_{\mathbf{x}} f_{\mathbf{X}}(\mathbf{x})|^2 = 0 \ .$$

Remark. In the proof of Theorem 5.1 we chose δ to be $4/51$, but the following proof shows that one can extend the range of δ beyond this value.

Proof. We shall show that

$$\int d\mathbf{X} \int_{\Omega_{\mathbf{X}}^c} d\mathbf{x} \, |\phi^{\mathrm{GP}}(\mathbf{x})|^2 |\nabla_{\mathbf{x}} f_{\mathbf{X}}(\mathbf{x})|^2$$

$$+ \int d\mathbf{X} \int d\mathbf{x} |\phi^{\mathrm{GP}}(\mathbf{x})|^2 |f_{\mathbf{X}}(\mathbf{x})|^2 \left[\frac{1}{2} \sum_{k\geq 2} v(|\mathbf{x} - \mathbf{x}_k|) - 8\pi N a |\phi^{\mathrm{GP}}(\mathbf{x})|^2 \right]$$

$$\geq -4\pi N a \int |\phi^{\mathrm{GP}}(\mathbf{x})|^4 d\mathbf{x} - o(1) \tag{7.8}$$

as $N \to \infty$. We claim that this implies the assertion of the Lemma. To see this, note that the left side of (7.8) can be written as

$$\frac{1}{N} E_0 - \mu^{\mathrm{GP}} - \int d\mathbf{X} \int_{\Omega_{\mathbf{X}}} d\mathbf{x} |\phi^{\mathrm{GP}}(\mathbf{x})|^2 |\nabla_{\mathbf{x}} f_{\mathbf{X}}(\mathbf{x})|^2 \ , \tag{7.9}$$

where we used partial integration and the GP equation (6.5), and also the symmetry of Ψ_0. The convergence of the energy in Theorem 6.1 and the relation (6.6) now imply the desired result.

The proof of (7.8) is actually just a detailed examination of the lower bounds to the energy derived in [LSeY1] and [LY1] and described in Chapters 2 and 6. We use the same methods as there, just describing the differences from the case considered here.

Writing

$$f_{\mathbf{x}}(\mathbf{x}) = \prod_{k \geq 2} \phi^{GP}(\mathbf{x}_k) F(\mathbf{x}, \mathbf{X}) \tag{7.10}$$

and using that F is symmetric in the particle coordinates, we see that (7.8) is equivalent to

$$\frac{1}{N} Q_\delta(F) \geq -4\pi N a \int |\phi^{GP}|^4 - o(1), \tag{7.11}$$

where Q_δ is the quadratic form

$$\begin{aligned} Q_\delta(F) &= \sum_{i=1}^{N} \int_{\Omega_i^c} |\nabla_i F|^2 \prod_{k=1}^{N} |\phi^{GP}(\mathbf{x}_k)|^2 d\mathbf{x}_k \\ &+ \sum_{1 \leq i < j \leq N} \int v(|\mathbf{x}_i - \mathbf{x}_j|) |F|^2 \prod_{k=1}^{N} |\phi^{GP}(\mathbf{x}_k)|^2 d\mathbf{x}_k \\ &- 8\pi N a \sum_{i=1}^{N} \int |\phi^{GP}(\mathbf{x}_i)|^2 |F|^2 \prod_{k=1}^{N} |\phi^{GP}(\mathbf{x}_k)|^2 d\mathbf{x}_k. \end{aligned} \tag{7.12}$$

Here Ω_i^c denotes the set

$$\Omega_i^c = \{(\mathbf{x}_1, \mathbf{X}) \in \mathbb{R}^{3N} | \min_{k \neq i} |\mathbf{x}_i - \mathbf{x}_k| \leq N^{-1/3-\delta}\}.$$

While (7.11) is not true for all conceivable F's satisfying the normalization condition

$$\int |F(\mathbf{x}, \mathbf{X})|^2 \prod_{k=1}^{N} |\phi^{GP}(\mathbf{x}_k)|^2 d\mathbf{x}_k = 1,$$

it *is* true for an F, such as ours, that has bounded kinetic energy (7.6). Looking at Chapter 6, we see that Eqs. (6.27)–(6.28), (6.47)–(6.51) are similar to (7.11), (7.12) and almost establish (7.11), but there are differences which we now explain.

In our case, the kinetic energy of particle i is restricted to the subset of $\mathbb{R}^{3N}$ in which $\min_{k \neq i} |\mathbf{x}_i - \mathbf{x}_k| \leq N^{-1/3-\delta}$. However, looking at the proof of the lower bound to the ground state energy of a homogeneous Bose gas discussed in Chapter 2, which enters the proof of Theorem 6.1, we see that if we choose $\delta \leq 4/51$ only this part of the kinetic energy is needed for the lower bound, except for some part with a relative magnitude of the order $\varepsilon = O(N^{-2\alpha})$ with $\alpha = 1/17$. (Here we use the a priori knowledge that the kinetic energy is bounded by (7.6).) We can even do better and choose some $4/51 < \delta < 2/9$, if α is chosen small enough. (To be precise, we choose $\beta = 1/3 + \alpha$ and $\gamma = 1/3 - 4\alpha$ in the notation of (2.69), and α small enough). The choice of α only affects the magnitude of the error term, however, which is still $o(1)$ as $N \to \infty$. $\square$

Proof of Theorem 7.1. For some $R > 0$ let $\mathcal{K} = \{\mathbf{x} \in \mathbb{R}^3, |\mathbf{x}| \leq R\}$, and define

$$\langle f\mathbf{x}\rangle_{\mathcal{K}} = \frac{1}{\int_{\mathcal{K}} |\phi^{\mathrm{GP}}(\mathbf{x})|^2 d\mathbf{x}} \int_{\mathcal{K}} |\phi^{\mathrm{GP}}(\mathbf{x})|^2 f\mathbf{x}(\mathbf{x})\, d\mathbf{x} \ .$$

We shall use Lemma 4.3, with $d = 3$, $h(\mathbf{x}) = |\phi^{\mathrm{GP}}(\mathbf{x})|^2 / \int_{\mathcal{K}} |\phi^{\mathrm{GP}}|^2$, $\Omega = \Omega_{\mathbf{X}} \cap \mathcal{K}$ and $f(\mathbf{x}) = f\mathbf{x}(\mathbf{x}) - \langle f\mathbf{x}\rangle_{\mathcal{K}}$ (see (7.7) and (7.5)). Since ϕ^{GP} is bounded on $\mathcal{K}$ above and below by some positive constants, this Lemma also holds (with a different constant C') with $d\mathbf{x}$ replaced by $|\phi^{\mathrm{GP}}(\mathbf{x})|^2 d\mathbf{x}$ in (4.4). Therefore,

$$\int d\mathbf{X} \int_{\mathcal{K}} d\mathbf{x} |\phi^{\mathrm{GP}}(\mathbf{x})|^2 \left[f\mathbf{x}(\mathbf{x}) - \langle f\mathbf{x}\rangle_{\mathcal{K}}\right]^2$$

$$\leq C' \int d\mathbf{X} \left[\int_{\Omega_{\mathbf{X}} \cap \mathcal{K}} |\phi^{\mathrm{GP}}(\mathbf{x})|^2 |\nabla_{\mathbf{x}} f\mathbf{x}(\mathbf{x})|^2 d\mathbf{x} \right.$$

$$\left. + \frac{N^{-2\delta}}{R^2} \int_{\mathcal{K}} |\phi^{\mathrm{GP}}(\mathbf{x})|^2 |\nabla_{\mathbf{x}} f\mathbf{x}(\mathbf{x})|^2 d\mathbf{x}\right], \tag{7.13}$$

where we used that $|\Omega^c_{\mathbf{X}} \cap \mathcal{K}| \leq (4\pi/3) N^{-3\delta}$. The first integral on the right side of (7.13) tends to zero as $N \to \infty$ by Lemma 7.3, and the second is bounded by (7.6). We conclude, since

$$\int_{\mathcal{K}} |\phi^{\mathrm{GP}}(\mathbf{x})|^2 f\mathbf{x}(\mathbf{x}) d\mathbf{x} \leq \int_{\mathbb{R}^3} |\phi^{\mathrm{GP}}(\mathbf{x})|^2 f\mathbf{x}(\mathbf{x}) d\mathbf{x}$$

because of the positivity of $f\mathbf{x}$, that

$$\liminf_{N\to\infty} \frac{1}{N} \langle \phi^{\mathrm{GP}} | \gamma | \phi^{\mathrm{GP}} \rangle \geq \int_{\mathcal{K}} |\phi^{\mathrm{GP}}(\mathbf{x})|^2 d\mathbf{x} \lim_{N\to\infty} \int d\mathbf{X} \int_{\mathcal{K}} d\mathbf{x} |\Psi_0(\mathbf{x}, \mathbf{X})|^2$$

$$= \left[\int_{\mathcal{K}} |\phi^{\mathrm{GP}}(\mathbf{x})|^2 d\mathbf{x}\right]^2,$$

where the last equality follows from (6.15). Since the radius of $\mathcal{K}$ was arbitrary, we conclude that

$$\lim_{N\to\infty} \frac{1}{N} \langle \phi^{\mathrm{GP}} | \gamma | \phi^{\mathrm{GP}} \rangle = 1,$$

implying convergence of γ/N to P^{GP} in Hilbert-Schmidt norm. Since the traces are equal, convergence even holds in trace norm (cf. [Si1], Thm. 2.20), and Theorem 7.1 is proved. □

We remark that the method presented here also works in the case of a two-dimensional Bose gas. The relevant parameter to be kept fixed in the GP limit is $N/|\ln(a^2\bar{\rho}_N)|$, all other considerations carry over without essential change, using the results in [LSeY2, LY2], c.f. Chapter 3 and Section 6.2. It should be noted that the existence of BEC in the ground state in 2D is not in conflict with its absence at positive temperatures [Ho, MW, M]. In the hard core lattice gas at half filling precisely this phenomenon occurs [KLS].

Finally, we remark on generalizations of Theorem 5.3 on superfluidity from a torus to some physically more realistic settings [LSeY5]. As an example, let $\mathcal{C}$ be a finite cylinder based on an annulus centered at the origin. Given a bounded, real function $a(r,z)$ let A be the vector field (in polar coordinates) $A(r,\theta,z) = \varphi a(r,z)\hat{e}_\theta$, where $\hat{e}_\theta$ is the unit vector in the θ direction. We also allow for a bounded external potential $V(r,z)$ that does not depend on θ.

Using the methods of Appendix A in [LSeY1], it is not difficult to see that there exists a $\varphi_0 > 0$, depending only on $\mathcal{C}$ and $a(r,z)$, such that for all $|\varphi| < \varphi_0$ there is a unique minimizer ϕ^{GP} of the Gross-Pitaevskii functional

$$\mathcal{E}^{\mathrm{GP}}[\phi] = \int_{\mathcal{C}} \left(|(\nabla + iA(\mathbf{x}))\phi(\mathbf{x})|^2 + V(\mathbf{x})|\phi(\mathbf{x})|^2 + 4\pi\mu Na|\phi(\mathbf{x})|^4 \right) d\mathbf{x} \quad (7.14)$$

under the normalization condition $\int |\phi|^2 = 1$. This minimizer does not depend on θ, and can be chosen to be positive, for the following reason: The relevant term in the kinetic energy is $T = -r^{-2}[\partial/\partial\theta + i\varphi r a(r,z)]^2$. If $|\varphi r a(r,z)| < 1/2$, it is easy to see that $T \geq \varphi^2 a(r,z)^2$, in which case, without raising the energy, we can replace ϕ by the square root of the θ-average of $|\phi|^2$. This can only lower the kinetic energy [LLo] and, by convexity of $x \to x^2$, this also lowers the ϕ^4 term.

We denote the ground state energy of $\mathcal{E}^{\mathrm{GP}}$ by E^{GP}, depending on Na and φ. The following Theorem 7.4 concerns the ground state energy E_0 of

$$H_N^A = \sum_{j=1}^{N} \left[-\left(\nabla_j + iA(\mathbf{x}_j)\right)^2 + V(\mathbf{x}_j) \right] + \sum_{1 \leq i < j \leq N} v(|\mathbf{x}_i - \mathbf{x}_j|) , \quad (7.15)$$

with Neumann boundary conditions on $\mathcal{C}$, and the one-particle reduced density matrix γ_N of the ground state, respectively. Different boundary conditions can be treated in the same manner, if they are also used in (7.14).

Remark. As a special case, consider a uniformly rotating system. In this case $A(\mathbf{x}) = \varphi r \widehat{e}_\theta$, where 2φ is the angular velocity. H_N^A is the Hamiltonian in the rotating frame, but with external potential $V(\mathbf{x}) + A(\mathbf{x})^2$ (see e.g. [Bm, p. 131]).

Theorem 7.4 (Superfluidity in a cylinder). *For $|\varphi| < \varphi_0$*

$$\lim_{N \to \infty} \frac{E_0(N, a, \varphi)}{N} = E^{\mathrm{GP}}(Na, \varphi) \tag{7.16}$$

in the limit $N \to \infty$ with Na fixed. In the same limit,

$$\lim_{N \to \infty} \frac{1}{N} \gamma_N(\mathbf{x}, \mathbf{x}') = \phi^{\mathrm{GP}}(\mathbf{x}) \phi^{\mathrm{GP}}(\mathbf{x}') \tag{7.17}$$

in trace class norm, i.e., $\lim_{N \to \infty} \mathrm{tr}\left[\,|\gamma_N/N - |\phi^{\mathrm{GP}}\rangle\langle\phi^{\mathrm{GP}}|\,|\,\right] = 0.$

Remark. In the special case of the curl-free vector potential $A(r, \theta) = \varphi r^{-1} \widehat{e}_\theta$, i.e., $a(r, z) = r^{-1}$, one can say more about the role of φ_0. In this case, there is a unique GP minimizer for all $\varphi \notin \mathbb{Z} + \frac{1}{2}$, whereas there are two minimizers for $\varphi \in \mathbb{Z} + \frac{1}{2}$. Part two of Theorem 7.4 holds in this special case for all $\varphi \notin \mathbb{Z} + \frac{1}{2}$, and (7.16) is true even for all φ.

In the case of a uniformly rotating system, where 2φ is the angular velocity, the condition $|\varphi| < \varphi_0$ in particular means that the angular velocity is smaller than the critical velocity for creating vortices [Se3, Se4, FS]. For rapidly rotating gases, the appearance of these vortices cause spontaneous breaking of the axial symmetry. The GP minimizer is then no longer unique, and Theorem 7.4 does not apply to this case. It *is*, however, possible to show that the GP equation still correctly describes the ground state of a dilute Bose gas in the rapidly rotating case, as was recently shown in [LSe2], using very different techniques than in the proof of Theorem 7.4.

Chapter 8

One-Dimensional Behavior of Dilute Bose Gases in Traps

Recently it has become possible to do experiments in highly elongated traps on ultra-cold Bose gases that are effectively one-dimensional [BBD, Go, G1, Sc, MSKE]. These experiments show peculiar features predicted by a model of a one-dimensional Bose gas with repulsive δ-function pair interaction, analyzed long ago by Lieb and Liniger [LL, L1].[1] These include quasi-fermionic behavior [Gi2], the absence of Bose-Einstein condensation (BEC) in a dilute limit [Le, PiSt, GWT], and an excitation spectrum different from that predicted by Bogoliubov's theory [L1, JK, KP]. The theoretical work on the dimensional cross-over for the ground state in elongated traps has so far been based either on variational calculations, starting from a 3D delta-potential [Ol, DGW, GW], or on numerical Quantum Monte Carlo studies [Bl, AG] with more realistic, genuine 3D potentials, but particle numbers limited to the order of 100. This work is important and has led to valuable insights, in particular about different parameter regions [PSW, DLO], but a more thorough theoretical understanding is clearly desirable since this is not a simple problem. In fact, it is evident that for a potential with a hard core the true 3D wave functions do not approximately factorize in the longitudinal and transverse variables (otherwise the energy would be infinite) and the effective 1D potential can not be obtained by simply integrating out the transverse variables of the 3D potential (that would immediately create an impenetrable barrier in 1D). It is important to be able to demonstrate rigorously, and therefore unambiguously,

---

[1]This model is discussed in Appendix B.

that the 1D behavior really follows from the fundamental Schrödinger equation. It is also important to delineate, as we do here, precisely what can be seen in the different parameter regions. The full proofs of our assertions are long and are given in [LSeY6]. Here we state our main results and outline the basic ideas for the proofs.

We start by describing the setting more precisely. It is convenient to write the Hamiltonian in the following way (in units where $\hbar = 2m = 1$):

$$H_{N,L,r,a} = \sum_{j=1}^{N} \left(-\Delta_j + V_r^{\perp}(\mathbf{x}_j^{\perp}) + V_L(z_j) \right) + \sum_{1 \leq i < j \leq N} v_a(|\mathbf{x}_i - \mathbf{x}_j|) \qquad (8.1)$$

with $\mathbf{x} = (x, y, z) = (\mathbf{x}^{\perp}, z)$ and with

$$V_r^{\perp}(\mathbf{x}^{\perp}) = \frac{1}{r^2} V^{\perp}(\mathbf{x}^{\perp}/r) ,$$

$$V_L(z) = \frac{1}{L^2} V(z/L) , \quad v_a(|\mathbf{x}|) = \frac{1}{a^2} v(|\mathbf{x}|/a) . \qquad (8.2)$$

Here, r, L, a are variable scaling parameters while $V^{\perp}, V$ and v are fixed.

We shall be concerned with the ground state of this Hamiltonian for large particle number N, which is appropriate for the consideration of actual experiments. The other parameters of the problem are the scattering length, a, of the two-body interaction potential, v, and two lengths, r and L, describing the transverse and the longitudinal extension of the trap potential, respectively.

The interaction potential v is supposed to be nonnegative, of finite range and have scattering length 1; the scaled potential v_a then has scattering length a. The external trap potentials V and $V^{\perp}$ confine the motion in the longitudinal (z) and the transversal ($\mathbf{x}^{\perp}$) directions, respectively, and are assumed to be continuous and tend to ∞ as $|z|$ and $|\mathbf{x}^{\perp}|$ tend to ∞. To simplify the discussion we find it also convenient to assume that V is homogeneous of some order $s > 0$, namely $V(z) = |z|^s$, but weaker assumptions, e.g. asymptotic homogeneity (cf. Chapter 6), would in fact suffice. The case of a simple box with hard walls is realized by taking $s = \infty$, while the usual harmonic approximation is $s = 2$. It is understood that the lengths associated with the ground states of $-d^2/dz^2 + V(z)$ and $-\Delta^{\perp} + V^{\perp}(\mathbf{x}^{\perp})$ are both of the order 1 so that L and r measure, respectively, the longitudinal and the transverse extensions of the trap. We denote the ground state energy of (8.1) by $E^{\mathrm{QM}}(N, L, r, a)$ and the ground state particle density by $\rho_{N,L,r,a}^{\mathrm{QM}}(\mathbf{x})$. On the average, this 3D density will always be low in the parameter range considered here (in the sense that distance between particles is large compared to the 3D scattering length). The effective 1D density can be either high or low, however.

In parallel with the 3D Hamiltonian we consider the Hamiltonian for n bosons in 1D with delta interaction and coupling constant $g \geq 0$, i.e.,

$$H_{n,g}^{1D} = \sum_{j=1}^{n} -\partial^2/\partial z_j^2 + g \sum_{1 \leq i < j \leq n} \delta(z_i - z_j) . \tag{8.3}$$

We consider this Hamiltonian for the z_j in an interval of length ℓ in the thermodynamic limit, $\ell \to \infty$, $n \to \infty$ with $\rho = n/\ell$ fixed. The ground state energy per particle in this limit is independent of boundary conditions and can, according to [LL] (see Appendix B), be written as

$$e_0^{1D}(\rho) = \rho^2 e(g/\rho) , \tag{8.4}$$

with a function $e(t)$ determined by a certain integral equation. Its asymptotic form is $e(t) \approx \frac{1}{2}t$ for $t \ll 1$ and $e(t) \to \pi^2/3$ for $t \to \infty$. Thus

$$e_0^{1D}(\rho) \approx \tfrac{1}{2}g\rho \quad \text{for} \quad g/\rho \ll 1 \tag{8.5}$$

and

$$e_0^{1D}(\rho) \approx (\pi^2/3)\rho^2 \quad \text{for} \quad g/\rho \gg 1 . \tag{8.6}$$

This latter energy is the same as for non-interacting fermions in 1D, which can be understood from the fact that (8.3) with $g = \infty$ is equivalent to a Hamiltonian describing free fermions.

Taking $\rho e_0^{1D}(\rho)$ as a local energy density for an inhomogeneous 1D system we can form the energy functional

$$\mathcal{E}[\rho] = \int_{-\infty}^{\infty} (|\nabla\sqrt{\rho}(z)|^2 + V_L(z)\rho(z) + \rho(z)^3 e(g/\rho(z))) \, dz . \tag{8.7}$$

Its ground state energy is obtained by minimizing over all normalized densities, i.e.,

$$E^{1D}(N, L, g) = \inf \left\{ \mathcal{E}[\rho] : \rho(z) \geq 0, \int_{-\infty}^{\infty} \rho(z)dz = N \right\}. \tag{8.8}$$

Using convexity of the map $\rho \mapsto \rho^3 e(g/\rho)$, it is standard to show that there exists a unique minimizer of (8.7) (see, e.g., [LSeY1]). It will be denoted by $\rho_{N,L,g}$. We also define the *mean 1D density* of this minimizer to be

$$\bar{\rho} = \frac{1}{N} \int_{-\infty}^{\infty} (\rho_{N,L,g}(z))^2 \, dz . \tag{8.9}$$

In a rigid box, i.e., for $s = \infty$, $\bar{\rho}$ is simply N/L (except for boundary corrections), but in more general traps it depends also on g besides N and L. The order of magnitude of $\bar{\rho}$ in the various parameter regions will be described below.

Our main result relates the 3D ground state energy of (8.1) to the 1D density functional energy $E^{1D}(N, L, g)$ in the large N limit with $g \sim a/r^2$ provided r/L and a/r are sufficiently small. To state this precisely, let $e^\perp$ and $b(\mathbf{x}^\perp)$, respectively, denote the ground state energy and the normalized ground state wave function of $-\Delta^\perp + V^\perp(\mathbf{x}^\perp)$. The corresponding quantities for $-\Delta^\perp + V_r^\perp(\mathbf{x}^\perp)$ are $e^\perp/r^2$ and $b_r(\mathbf{x}^\perp) = (1/r)b(\mathbf{x}^\perp/r)$. In the case that the trap is a cylinder with hard walls b is a Bessel function; for a quadratic $V^\perp$ it is a Gaussian.

Define g by

$$g = \frac{8\pi a}{r^2} \int |b(\mathbf{x}^\perp)|^4 d\mathbf{x}^\perp = 8\pi a \int |b_r(\mathbf{x}^\perp)|^4 d\mathbf{x}^\perp. \tag{8.10}$$

Our main result of this chapter is:

Theorem 8.1 (From 3D to 1D). *Let $N \to \infty$ and simultaneously $r/L \to 0$ and $a/r \to 0$ in such a way that $r^2 \bar\rho \cdot \min\{\bar\rho, g\} \to 0$. Then*

$$\lim \frac{E^{QM}(N, L, r, a) - Ne^\perp/r^2}{E^{1D}(N, L, g)} = 1. \tag{8.11}$$

An analogous result hold for the ground state density. Define the 1D QM density by averaging over the transverse variables, i.e.,

$$\hat\rho_{N,L,r,a}^{QM}(z) \equiv \int \rho_{N,L,r,a}^{QM}(\mathbf{x}^\perp, z)d\mathbf{x}^\perp . \tag{8.12}$$

Let $\bar L := N/\bar\rho$ denote the extension of the system in z-direction, and define the rescaled density $\tilde\rho$ by

$$\rho_{N,L,g}^{1D}(z) = \frac{N}{\bar L}\tilde\rho(z/\bar L) . \tag{8.13}$$

Note that, although $\tilde\rho$ depends on N, L and g, $\|\tilde\rho\|_1 = \|\tilde\rho\|_2 = 1$, which shows in particular that $\bar L$ is the relevant scale in z-direction. The result for the ground state density is:

Theorem 8.2 (1D limit for density). *In the same limit as considered in Theorem 8.1,*

$$\lim \left(\frac{\bar L}{N}\hat\rho_{N,L,r,a}^{QM}(z\bar L) - \tilde\rho(z) \right) = 0 \tag{8.14}$$

in weak L^1 sense.

Note that because of (8.5) and (8.6) the condition $r^2\bar\rho \cdot \min\{\bar\rho, g\} \to 0$ is the same as

$$e_0^{1D}(\bar\rho) \ll 1/r^2 , \tag{8.15}$$

i.e., the average energy per particle associated with the longitudinal motion should be much smaller than the energy gap between the ground and first excited state of the confining Hamiltonian in the transverse directions. Thus, the basic physics is highly quantum-mechanical and has no classical counterpart. The system can be described by a 1D functional (8.7), *even though the transverse trap dimension is much larger than the range of the atomic forces.*

8.1 Discussion of the Results

We will now give a discussion of the various parameter regions that are included in the limit considered in Theorems 8.1 and 8.2 above. We begin by describing the division of the space of parameters into two basic regions. This decomposition will eventually be refined into five regions, but for the moment let us concentrate on the basic dichotomy.

In Chapter 6 we proved that the 3D Gross-Pitaevskii formula for the energy is correct to leading order in situations in which N is large but a is small compared to the mean particle distance. This energy has two parts: The energy necessary to confine the particles in the trap, plus the internal energy of interaction, which is $N4\pi a\rho^{3D}$. This formula was proved to be correct for a *fixed* confining potential in the limit $N \to \infty$ with $a^3\rho^{3D} \to 0$. However, this limit does not hold uniformly if r/L gets small as N gets large. In other words, new physics can come into play as $r/L \to 0$ and it turns out that this depends on the ratio of a/r^2 to the 1D density, or, in other words, on $g/\bar{\rho}$. There are two basic regimes to consider in highly elongated traps, i.e., when $r \ll L$. They are

- The 1D limit of the 3D Gross-Pitaevskii regime

- The 'true' 1D regime.

The former is characterized by $g/\bar{\rho} \ll 1$, while in the latter regime $g/\bar{\rho}$ is of the order one or even tends to infinity. (If $g/\bar{\rho} \to \infty$ the particles are effectively impenetrable; this is usually referred to as the Girardeau-Tonks region.) These two situations correspond to high 1D density (weak interaction) and low 1D density (strong interaction), respectively. Physically, the main difference is that in the strong interaction regime the motion of the particles in the longitudinal direction is highly correlated, while in the weak interaction regime it is not. Mathematically, this distinction also shows up in our proofs. The first region is correctly described by both the 3D and 1D theories because the two give the same predictions there. That's why we call the second region the 'true' 1D regime.

In both regions the internal energy of the gas is small compared to the energy of confinement. However, this in itself does not imply a specifically 1D behavior. (If a is sufficiently small it is satisfied in a trap of any shape.) 1D behavior, when it occurs, manifests itself by the fact that the transverse motion of the atoms is uncorrelated while the longitudinal motion is correlated (very roughly speaking) in the same way as pearls on a necklace. Thus, the true criterion for 1D behavior is that $g/\bar{\rho}$ is of order unity or larger and not merely the condition that the energy of confinement dominates the internal energy.

We shall now briefly describe the finer division of these two regimes into five regions altogether. Three of them (Regions 1–3) belong to the weak interaction regime and two (Regions 4–5) to the strong interaction regime. They are characterized by the behavior of $g/\bar{\rho}$ as $N \to \infty$. In each of these regions the general functional (8.7) can be replaced by a different, simpler functional, and the energy $E^{1D}(N, L, g)$ in Theorem 8.1 by the ground state energy of that functional. Analogously, the density in Theorem 8.2 can be replaced by the minimizer of the functional corresponding to the region considered.

The five regions are

• **Region 1, the Ideal Gas case:** In the trivial case where the interaction is so weak that it effectively vanishes in the large N limit and everything collapses to the ground state of $-d^2/dz^2 + V(z)$ with ground state energy $e^{\parallel}$, the energy E^{1D} in (8.11) can be replaced by $Ne^{\parallel}/L^2$. This is the case if $g/\bar{\rho} \ll N^{-2}$, and the mean density is just $\bar{\rho} \sim N/L$. Note that $g/\bar{\rho} \ll N^{-2}$ means that the 3D interaction energy per particle $\sim a\rho^{3D} \ll 1/L^2$.

• **Region 2, the 1D GP case:** In this region $g/\bar{\rho} \sim N^{-2}$, with $\bar{\rho} \sim N/L$. This case is described by a 1D Gross-Pitaevskii energy functional of the form

$$\mathcal{E}_{1D}^{GP}[\rho] = \int_{-\infty}^{\infty} \left(|\nabla\sqrt{\rho(z)}|^2 + V_L(z)\rho(z) + \tfrac{1}{2}g\rho(z)^2 \right) dz \ , \tag{8.16}$$

corresponding to the high density approximation (8.5) of the interaction energy in (8.7). Its ground state energy, E_{1D}^{GP}, fulfills the scaling relation $E_{1D}^{GP}(N, L, g) = NL^{-2}E_{1D}^{GP}(1, 1, NgL)$.

• **Region 3, the 1D TF case:** $N^{-2} \ll g/\bar{\rho} \ll 1$, with $\bar{\rho}$ being of the order $\bar{\rho} \sim (N/L)(NgL)^{-1/(s+1)}$, where s is the degree of homogeneity of the longitudinal confining potential V. This region is described by a Thomas-Fermi type functional

$$\mathcal{E}_{1D}^{TF}[\rho] = \int_{-\infty}^{\infty} \left(V_L(z)\rho(z) + \tfrac{1}{2}g\rho(z)^2 \right) dz \ . \tag{8.17}$$

It is a limiting case of Region 2 in the sense that $NgL \gg 1$, but a/r is sufficiently small so that $g/\bar{\rho} \ll 1$, i.e., the high density approximation in (8.5) is still valid. The explanation of the factor $(NgL)^{1/(s+1)}$ is as follows: The linear extension $\bar{L}$ of the minimizing density of (8.16) is for large values of NgL determined by $V_L(\bar{L}) \sim g(N/\bar{L})$, which gives $\bar{L} \sim (NgL)^{1/(s+1)}L$. In addition condition (8.15) requires $g\bar{\rho} \ll r^{-2}$, which means that $Na/L(NgL)^{1/(s+1)} \ll 1$. The minimum energy of (8.17) has the scaling property $E_{1D}^{TF}(N, L, g) = NL^{-2}(NgL)^{s/(s+1)}E_{1D}^{TF}(1, 1, 1)$.

- **Region 4, the LL case:** $g/\bar{\rho} \sim 1$, with $\bar{\rho} \sim (N/L)N^{-2/(s+2)}$, described by an energy functional

$$\mathcal{E}^{LL}[\rho] = \int_{-\infty}^{\infty} \left(V_L(z)\rho(z) + \rho(z)^3 e(g/\rho(z)) \right) dz \ . \tag{8.18}$$

This region corresponds to the case $g/\bar{\rho} \sim 1$, so that neither the high density (8.5) nor the low density approximation (8.6) is valid and the full LL energy (8.4) has to be used. The extension $\bar{L}$ of the system is now determined by $V_L(\bar{L}) \sim (N/\bar{L})^2$ which leads to $\bar{L} \sim LN^{2/(s+2)}$. Condition (8.15) means in this region that $Nr/\bar{L} \sim N^{s/(s+2)}r/L \to 0$. Since $Nr/\bar{L} \sim (\bar{\rho}/g)(a/r)$, this condition is automatically fulfilled if $g/\bar{\rho}$ is bounded away from zero and $a/r \to 0$. The ground state energy of (8.18), $E^{LL}(N, L, g)$, is equal to $N\gamma^2 E^{LL}(1, 1, g/\gamma)$, where we introduced the density parameter $\gamma := (N/L)N^{-2/(s+2)}$.

- **Region 5, the GT case:** $g/\bar{\rho} \gg 1$, with $\bar{\rho} \sim (N/L)N^{-2/(s+2)}$, described by a functional with energy density $\sim \rho^3$, corresponding to the Girardeau-Tonks limit of the LL energy density. It corresponds to impenetrable particles, i.e, the limiting case $g/\bar{\rho} \to \infty$ and hence formula (8.6) for the energy density. As in Region 4, the mean density is here $\bar{\rho} \sim \gamma$. The energy functional is

$$\mathcal{E}^{GT}[\rho] = \int_{-\infty}^{\infty} \left(V_L(z)\rho(z) + (\pi^2/3)\rho(z)^3 \right) dz \ , \tag{8.19}$$

with minimum energy $E^{GT}(N, L) = N\gamma^2 E^{GT}(1, 1)$.

As already mentioned above, Regions 1–3 can be reached as limiting cases of a 3D Gross-Pitaevskii theory. In this sense, the behavior in these regions contains remnants of the 3D theory, which also shows up in the fact that BEC prevails in Regions 1 and 2 (See [LSeY6] for details.) Heuristically, these traces of 3D can be understood from the fact that in Regions 1–3 the 1D formula for energy per particle, $g\rho \sim aN/(r^2L)$, gives the same result as the 3D formula, i.e., scattering length times 3D density. This is no longer so in Regions 4 and 5 and different methods are required.

8.2   The 1D Limit of 3D GP Theory

Before discussing the many-body problem we treat the much simpler problem of taking the $r/L \to 0$ limit of the 3D GP ground state energy $E_{3D}^{GP}(N, L, r, a)$, defined by (6.4) with the potential $V(\mathbf{x}) = V_r^\perp(\mathbf{x}^\perp) + V_L(z)$. The result is, apart from the confining energy, the 1D GP energy with coupling constant $g \sim a/r^2$. In particular we see that Regions 4 and 5 cannot be reached as a limit of 3D GP theory.

Theorem 8.3 (1D limit of 3D GP energy). *Let g be given by (8.10). If $r/L \to 0$, then*

$$\frac{E_{3D}^{GP}(N, L, r, a) - Nr^{-2}e^\perp}{E_{1D}^{GP}(N, L, g)} \to 1 \qquad (8.20)$$

uniformly in the parameters, as long as $\bar{\rho}a \to 0$.

Remark. Since $E^{GP}(1, L, Ng)_{1D} \sim L^{-2} + \bar{\rho}a/r^2$, the condition $\bar{\rho}a \to 0$ is equivalent to $r^2 E_{1D}^{GP}(1, L, Ng) \to 0$, which means simply that the 1D GP energy per particle is much less than the confining energy, $\sim 1/r^2$.

Proof. Because of the scaling relation (8.16) and the corresponding relation for E_{3D}^{GP} it suffices to consider the case $N = 1$ and $L = 1$.

We denote the (positive) minimizer of the one-dimensional GP functional (8.16) with $N = 1$, $L = 1$ and g fixed by $\phi(z)$. Taking $b_r(\mathbf{x}^\perp)\phi(z)$ as trial function for the 3D functional (6.2) and using the definition (8.10) of g we obtain without further ado the upper bound

$$E_{3D}^{GP}(1, 1, r, a) \leq e^\perp/r^2 + E_{1D}^{GP}(1, 1, g) \qquad (8.21)$$

for all r and a.

For a lower bound we consider the one-particle Hamiltonian

$$H_{r,a} = -\Delta^\perp + V_r^\perp(\mathbf{x}^\perp) - \partial_z^2 + V(z) + 8\pi a b_r(\mathbf{x}^\perp)^2 \phi(z)^2 . \qquad (8.22)$$

Taking the 3D Gross-Pitaevskii minimizer $\Phi(\mathbf{x})$ for $N = 1$, $L = 1$, as trial function we get

$$
\begin{aligned}
\inf \operatorname{spec} H_{r,a} \quad &\leq \quad E_{3D}^{GP}(1, 1, r, a) - 4\pi a \int \Phi^4 + 8\pi a \int b_r^2 \phi^2 \Phi^2 \\
&\leq \quad E_{3D}^{GP}(1, 1, r, a) + 4\pi a \int b_r^4 \phi^4 \\
&= \quad E_{3D}^{GP}(1, 1, r, a) + \frac{g}{2} \int \phi^4 . \qquad (8.23)
\end{aligned}
$$

To bound $H_{r,a}$ from below we consider first for fixed $z \in \mathbb{R}$ the Hamiltonian (in 2D)

$$H_{r,a,z} = -\Delta^\perp + V_r(\mathbf{x}^\perp) + 8\pi a\phi(z)^2 b_r(\mathbf{x}^\perp)^2. \tag{8.24}$$

We regard $-\Delta^\perp + V_r(\mathbf{x}^\perp)$ as its "free" part and $8\pi a\phi(z)^2 b_r(\mathbf{x}^\perp)^2$ as a perturbation. Since the perturbation is positive all the eigenvalues of $H_{r,a,z}$ are at least as large as those of $-\Delta^\perp + V_r(\mathbf{x}^\perp)$; in particular, the first excited eigenvalue is $\sim 1/r^2$. The expectation value in the ground state b_r of the free part is

$$\langle H_{r,a,z} \rangle = e^\perp/r^2 + g|\phi(z)|^2. \tag{8.25}$$

Temple's inequality (2.51) gives

$$H_{r,a,z} \geq \left(e^\perp/r^2 + g|\phi(z)|^2 \right) \left(1 - \frac{\langle (H_{r,a,z} - \langle H_{r,a,z} \rangle)^2 \rangle}{\langle H_{r,a,z} \rangle (\tilde{e}^\perp/r^2 - \langle H_{r,a,z} \rangle)} \right) \tag{8.26}$$

where $\tilde{e}^\perp/r^2$ is the lowest eigenvalue above the ground state energy of $-\Delta^\perp + V_r(\mathbf{x}^\perp)$. Since

$$H_{r,a,z} b_r = (e^\perp/r^2) b_r + 8\pi a\phi(z)^2 b_r^3 \tag{8.27}$$

we have $(H_{r,a,z} - \langle H_{r,a,z} \rangle) b_r = 8\pi \phi(z)^2 a b_r^3 - g\phi(z)^2 b_r)$ and hence, using that $g = 8\pi a \int b_r^4 = 8\pi(a/r^2) \int b^4$,

$$
\begin{aligned}
\langle (H_{r,a,z} - \langle H_{r,a,z} \rangle)^2 \rangle &= \phi(z)^4 \int \left(8\pi a b_r(\mathbf{x}^\perp)^3 - g b_r(\mathbf{x}^\perp) \right)^2 d\mathbf{x}^\perp \\
&= \phi(z)^4 \int \left[(8\pi a)^2 b_r(\mathbf{x}^\perp)^6 - 16\pi a g b_r(\mathbf{x}^\perp)^4 \right. \\
&\quad \left. + g^2 b_r(\mathbf{x}^\perp)^2 \right] d\mathbf{x}^\perp \\
&\leq \text{const.} \|\phi\|_\infty^4 g^2 \leq \text{const.} E_{1D}^{\mathrm{GP}}(1,1,g)^2, \tag{8.28}
\end{aligned}
$$

where we have used Lemma 2.1 in [LSeY2] to bound $g\|\phi\|_\infty^2$ by const. $E_{1D}^{\mathrm{GP}}(1,1,g)$. We thus see from (8.25) and the assumption $r^2 E_{1D}^{\mathrm{GP}}(1,1,g) \to 0$ that the error term in the Temple inequality (8.26) is $o(1)$.

Now $H_{r,a} = -\partial_z^2 + V(z) + H_{r,a,z}$, so from (8.26) we conclude that

$$H_{r,a} \geq \left((e^\perp/r^2) - \partial_z^2 + V(z) + g|\phi(x)|^2 \right)(1 - o(1)). \tag{8.29}$$

On the other hand, the lowest energy of $-\partial_z^2 + V(z) + g|\phi(z)|^2$ is just $E_{1D}^{\mathrm{GP}}(1,1,g) + (g/2) \int_{\mathbb{R}} |\phi(z)|^4 dz$. Combining (8.23) and (8.26) we thus get

$$E_{3D}^{\mathrm{GP}}(1,1,r,a) - e^\perp/r^2 \geq E_{1D}^{\mathrm{GP}}(1,1,g)(1 - o(1)). \tag{8.30}$$

$\square$

8.3   Outline of Proof

We now outline the main steps in the proof of Theorems 8.1 and 8.2, referring to [LSeY6] for full details. To prove (8.11) one has to establish upper and lower bounds, with controlled errors, on the QM many-body energy in terms of the energies obtained by minimizing the energy functionals appropriate for the various regions. The limit theorem for the densities can be derived from the energy estimates in a standard way by variation with respect to the external potential V_L.

The different parameter regions have to be treated by different methods, a watershed lying between Regions 1–3 on the one hand and Regions 4–5 on the other. In Regions 1–3, similar methods as in the proof of the 3D Gross-Pitaevskii limit theorem discussed in Chapter 6 can be used. This 3D proof needs some modifications, however, because there the external potential was fixed and the estimates are not uniform in the ratio r/L. We will not go into the details here, but mainly focus on Regions 4 and 5, where new methods are needed. It turns out to be necessary to localize the particles by dividing the trap into finite 'boxes' (finite in z-direction), with a controllable particle number in each box. The particles are then distributed optimally among the boxes to minimize the energy, in a similar way as Eq. (2.65) was derived from Eq. (2.60).

A core lemma for Regions 4–5 is an estimate of the 3D ground state energy in a finite box in terms of the 1D energy of the Hamiltonian (8.3). I.e., we will consider the ground state energy of (8.1) with the external potential $V_L(z)$ replaced by a finite box (in z-direction) with length ℓ. Let $E_{\mathrm{D}}^{\mathrm{QM}}(n,\ell,r,a)$ and $E_{\mathrm{N}}^{\mathrm{QM}}(n,\ell,r,a)$ denote its ground state energy with Dirichlet and Neumann boundary conditions, respectively.

Lemma 8.4. *Let $E_{\mathrm{D}}^{\mathrm{1D}}(n,\ell,g)$ and $E_{\mathrm{N}}^{\mathrm{1D}}(n,\ell,g)$ denote the ground state energy of (8.3) on $L^2([0,\ell]^n)$, with Dirichlet and Neumann boundary conditions, respectively, and let g be given by (8.10). Then there is a finite number $C > 0$ such that*

$$E_{\mathrm{N}}^{\mathrm{QM}}(n,\ell,r,a) - \frac{ne^{\perp}}{r^2} \geq E_{\mathrm{N}}^{\mathrm{1D}}(n,\ell,g)\left(1 - Cn\left(\frac{a}{r}\right)^{1/8}\left[1 + \frac{nr}{\ell}\left(\frac{a}{r}\right)^{1/8}\right]\right) . \quad (8.31)$$

Moreover,

$$E_{\mathrm{D}}^{\mathrm{QM}}(n,\ell,r,a) - \frac{ne^{\perp}}{r^2} \leq E_{\mathrm{D}}^{\mathrm{1D}}(n,\ell,g)\left(1 + C\left[\left(\frac{na}{r}\right)^2\left(1 + \frac{a\ell}{r^2}\right)\right]^{1/3}\right) , \quad (8.32)$$

provided the term in square brackets is less than 1.

This Lemma is the key to the proof of Theorems 8.1 and 8.2. The reader interested in the details is referred to [LSeY6]. Here we only give a sketch of the proof of Lemma 8.4.

Proof of Lemma 8.4. We start with the upper bound (8.32). Let ψ denote the ground state of (8.3) with Dirichlet boundary conditions, normalized by $\langle\psi|\psi\rangle = 1$, and let $\rho_\psi^{(2)}$ denote its two-particle density, normalized by $\int \rho_\psi^{(2)}(z,z')dzdz' = 1$. Let G and F be given by $G(\mathbf{x}_1,\ldots,\mathbf{x}_n) = \psi(z_1,\ldots,z_n)\prod_{j=1}^n b_r(\mathbf{x}_j^\perp)$ and by $F(\mathbf{x}_1,\ldots,\mathbf{x}_n) = \prod_{i<j} f(|\mathbf{x}_i - \mathbf{x}_j|)$. Here f is a monotone increasing function, with $0 \le f \le 1$ and $f(t) = 1$ for $t \ge R$ for some $R \ge R_0$. For $t \le R$ we shall choose $f(t) = f_0(t)/f_0(R)$, where f_0 is the solution to the zero-energy scattering equation for v_a (2.3). Note that $f_0(R) = 1 - a/R$ for $R \ge R_0$, and $f_0'(t) \le t^{-1}\min\{1, a/t\}$. We use as a trial wave function

$$\Psi(\mathbf{x}_1,\ldots,\mathbf{x}_n) = G(\mathbf{x}_1,\ldots,\mathbf{x}_n)F(\mathbf{x}_1,\ldots,\mathbf{x}_n) . \tag{8.33}$$

We first have to estimate the norm of Ψ. Using the fact that F is 1 whenever no pair of particles is closer together than a distance R, we obtain

$$\langle\Psi|\Psi\rangle \ge 1 - \frac{n(n-1)}{2}\frac{\pi R^2}{r^2}\|b\|_4^4 . \tag{8.34}$$

To evaluate the expectation value of the Hamiltonian, we use

$$\langle\Psi| - \Delta_j|\Psi\rangle = -\int F^2 G\Delta_j G + \int G^2|\nabla_j F|^2 \tag{8.35}$$

and the Schrödinger equation $H_{n,g}\psi = E_D^{1D}\psi$. This gives

$$\langle\Psi|H|\Psi\rangle = \left(E_D^{1D} + \frac{n}{r^2}e^\perp\right)\langle\Psi|\Psi\rangle - g\langle\Psi|\sum_{i<j}\delta(z_i - z_j)|\Psi\rangle$$

$$+ \int G^2\left(\sum_{j=1}^n|\nabla_j F|^2 + \sum_{i<j}v_a(|\mathbf{x}_i - \mathbf{x}_j|)|F|^2\right) . \tag{8.36}$$

Now, since $0 \le f \le 1$ and $f' \ge 0$ by assumption, $F^2 \le f(|\mathbf{x}_i - \mathbf{x}_j|)^2$, and

$$\sum_{j=1}^n|\nabla_j F|^2 \le 2\sum_{i<j}f'(|\mathbf{x}_i - \mathbf{x}_j|)^2 + 4\sum_{k<i<j}f'(|\mathbf{x}_k - \mathbf{x}_i|)f'(|\mathbf{x}_k - \mathbf{x}_j|) . \tag{8.37}$$

Consider the first term on the right side of (8.37), together with the last term in (8.36). These terms are bounded above by

$$n(n-1)\int b_r(\mathbf{x}^\perp)^2 b_r(\mathbf{y}^\perp)^2\rho_\psi^{(2)}(z,z')\left(f'(|\mathbf{x} - \mathbf{y}|)^2 + \tfrac{1}{2}v_a(|\mathbf{x} - \mathbf{y}|)f(|\mathbf{x} - \mathbf{y}|)^2\right) . \tag{8.38}$$

Let

$$h(z) = \int \left(f'(|\mathbf{x}|)^2 + \tfrac{1}{2}v_a(|\mathbf{x}|)f(|\mathbf{x}|)^2 \right) d\mathbf{x}^\perp . \tag{8.39}$$

Using Young's inequality for the integration over the $\perp$-variables, we get

$$(8.38) \leq \frac{n(n-1)}{r^2} \|b\|_4^4 \int_{\mathbb{R}^2} \rho_\psi^{(2)}(z, z') h(z - z') dz dz' . \tag{8.40}$$

By similar methods, one can show that the contribution from the last term in (8.37) is bounded by

$$\frac{2}{3}n(n-1)(n-2)\frac{\|b\|_\infty^2}{r^2} \frac{\|b\|_4^4}{r^2} \|k\|_\infty \int_{\mathbb{R}^2} \rho_\psi^{(2)}(z, z') k(z - z') dz dz' , \tag{8.41}$$

where

$$k(z) = \int f'(|\mathbf{x}|) d\mathbf{x}^\perp . \tag{8.42}$$

Note that both h and k are supported in $[-R, R]$.

Now, for any $\phi \in H^1(\mathbb{R})$,

$$\left| |\phi(z)|^2 - |\phi(z')|^2 \right| \leq 2|z - z'|^{1/2} \left(\int_\mathbb{R} |\phi|^2 \right)^{1/4} \left(\int_\mathbb{R} \left| \frac{d\phi}{dz} \right|^2 \right)^{3/4} . \tag{8.43}$$

Applying this to $\rho_\psi^{(2)}(z, z')$, considered as a function of z only, we get

$$\int_{\mathbb{R}^2} \rho_\psi^{(2)}(z, z') h(z - z') dz dz' - \int_\mathbb{R} h(z) dz \int \rho_\psi^{(2)}(z, z) dz$$
$$\leq 2R^{1/2} \int_\mathbb{R} h(z) dz \left\langle \psi \left| -\frac{d^2}{dz_1^2} \right| \psi \right\rangle^{3/4} , \tag{8.44}$$

where we used Schwarz's inequality, the normalization of $\rho_\psi^{(2)}$ and the symmetry of ψ. The same argument is used for (8.41) with h replaced by k.

It remains to bound the second term in (8.36). As in the estimate for the norm of Ψ, we use again the fact that F is equal to 1 as long as the particles are not within a distance R from each other. We obtain

$$\langle \Psi | \sum_{i<j} \delta(z_i - z_j) | \Psi \rangle \geq \frac{n(n-1)}{2} \int \rho_\psi^{(2)}(z, z) dz \left(1 - \frac{n(n-1)}{2} \frac{\pi R^2}{r^2} \|b\|_4^4 \right) . \tag{8.45}$$

We also estimate $g\frac{1}{2}n(n-1) \int \rho_\psi^{(2)}(z, z) dz \leq E_D^{1D}$ and $\langle \psi | - d^2/dz_1^2 | \psi \rangle \leq E_D^{1D}/n$. We have $\int h(z) dz = 4\pi a(1 - a/R)^{-1}$, and the terms containing k can be bounded by $\|k\|_\infty \leq 2\pi a(1+\ln(R/a))/(1-a/r)$ and $\int k(z)dz \leq 2\pi aR(1-a/(2R))/(1-a/r)$. Putting together all the bounds, and choosing

$$R^3 = \frac{ar^2}{n^2(1+g\ell)} , \tag{8.46}$$

this proves the desired result.

We are left with the lower bound (8.31). We write a general wave function Ψ as

$$\Psi(\mathbf{x}_1,\ldots,\mathbf{x}_n) = f(\mathbf{x}_1,\ldots,\mathbf{x}_n) \prod_{k=1}^{n} b_r(\mathbf{x}_k^\perp) , \tag{8.47}$$

which can always be done, since b_r is a strictly positive function. Partial integration gives

$$\langle \Psi|H|\Psi \rangle = \frac{ne^\perp}{r^2} + \sum_{i=1}^{n} \int \left[|\nabla_i f|^2 + \tfrac{1}{2} \sum_{j,\,j\neq i} v_a(|\mathbf{x}_i - \mathbf{x}_j|)|f|^2 \right] \prod_{k=1}^{n} b_r(\mathbf{x}_k^\perp)^2 d\mathbf{x}_k . \tag{8.48}$$

Choose some $R > R_0$, fix i and $\mathbf{x}_j$, $j \neq i$, and consider the Voronoi cell Ω_j around particle j, i.e., $\Omega_j = \{\mathbf{x} : |\mathbf{x} - \mathbf{x}_j| \leq |\mathbf{x} - \mathbf{x}_k| \text{ for all } k \neq j\}$. If $\mathcal{B}_j$ denotes the ball of radius R around $\mathbf{x}_j$, we can estimate with the aid of Lemma 2.5

$$\int_{\Omega_j \cap \mathcal{B}_j} b_r(\mathbf{x}_i^\perp)^2 \left(|\nabla_i f|^2 + \tfrac{1}{2} v_a(|\mathbf{x}_i - \mathbf{x}_j|)|f|^2 \right) d\mathbf{x}_i$$

$$\geq \frac{\min_{\mathbf{x}\in\mathcal{B}_j} b_r(\mathbf{x}^\perp)^2}{\max_{\mathbf{x}\in\mathcal{B}_j} b_r(\mathbf{x}^\perp)^2} a \int_{\Omega_j \cap \mathcal{B}_j} b_r(\mathbf{x}_i^\perp)^2 U(|\mathbf{x}_i - \mathbf{x}_j|)|f|^2 . \tag{8.49}$$

Here U is given in (2.43). For some $\delta > 0$ let $\mathcal{B}_\delta$ be the subset of $\mathbb{R}^2$ where $b(\mathbf{x}^\perp)^2 \geq \delta$, and let $\chi_{\mathcal{B}_\delta}$ denote its characteristic function. Estimating $\max_{\mathbf{x}\in\mathcal{B}_j} b_r(\mathbf{x}^\perp)^2 \leq \min_{\mathbf{x}\in\mathcal{B}_j} b_r(\mathbf{x}^\perp)^2 + 2(R/r^3)\|\nabla b^2\|_\infty$, we obtain

$$\frac{\min_{\mathbf{x}\in\mathcal{B}_j} b_r(\mathbf{x}^\perp)^2}{\max_{\mathbf{x}\in\mathcal{B}_j} b_r(\mathbf{x}^\perp)^2} \geq \chi_{\mathcal{B}_\delta}(\mathbf{x}_j^\perp/r) \left(1 - 2\frac{R}{r}\frac{\|\nabla b^2\|_\infty}{\delta} \right) . \tag{8.50}$$

Denoting $k(i)$ the nearest neighbor to particle i, we conclude that, for $0 \leq \varepsilon \leq 1$,

$$\sum_{i=1}^{n} \int \left[|\nabla_i f|^2 + \tfrac{1}{2} \sum_{j,\,j\neq i} v_a(|\mathbf{x}_i - \mathbf{x}_j|)|f|^2 \right] \prod_{k=1}^{n} b_r(\mathbf{x}_k^\perp)^2 d\mathbf{x}_k$$

$$\geq \sum_{i=1}^{n} \int \left[\varepsilon|\nabla_i f|^2 + (1-\varepsilon)|\nabla_i f|^2 \chi_{\min_k |z_i - z_k| \geq R}(z_i) \right.$$

$$\left. + a'U(|\mathbf{x}_i - \mathbf{x}_{k(i)}|)\chi_{\mathcal{B}_\delta}(\mathbf{x}_{k(i)}^\perp/r)|f|^2 \right] \prod_{k=1}^{n} b_r(\mathbf{x}_k^\perp)^2 d\mathbf{x}_k , \tag{8.51}$$

where $a' = a(1 - \varepsilon)(1 - 2R\|\nabla b^2\|_\infty/r\delta)$.

Define $F(z_1, \ldots, z_n) \geq 0$ by

$$|F(z_1, \ldots, z_n)|^2 = \int |f(\mathbf{x}_1, \ldots, \mathbf{x}_n)|^2 \prod_{k=1}^{n} b_r(\mathbf{x}_k^\perp)^2 d\mathbf{x}_k^\perp . \tag{8.52}$$

Neglecting the kinetic energy in $\perp$-direction in the second term in (8.51) and using the Schwarz inequality to bound the longitudinal kinetic energy of f by the one of F, we get the estimate

$$\langle \Psi | H | \Psi \rangle - \frac{ne^\perp}{r^2} \geq$$

$$\sum_{i=1}^{n} \int \left[\varepsilon |\partial_i F|^2 + (1 - \varepsilon) |\partial_i F|^2 \chi_{\min_k |z_i - z_k| \geq R}(z_i) \right] \prod_{k=1}^{n} dz_k$$

$$+ \sum_{i=1}^{n} \int \left[\varepsilon |\nabla_i^\perp f|^2 + a'U(|\mathbf{x}_i - \mathbf{x}_{k(i)}|)\chi_{\mathcal{B}_\delta}(\mathbf{x}_{k(i)}^\perp/r)|f|^2 \right] \prod_{k=1}^{n} b_r(\mathbf{x}_k^\perp)^2 d\mathbf{x}_k ,$$

$$\tag{8.53}$$

where $\partial_j = d/dz_j$, and $\nabla^\perp$ denotes the gradient in $\perp$-direction. We now investigate the last term in (8.53). Consider, for fixed $z_1, \ldots, z_n$, the expression

$$\sum_{i=1}^{n} \int \left[\varepsilon |\nabla_i^\perp f|^2 + a'U(|\mathbf{x}_i - \mathbf{x}_{k(i)}|)\chi_{\mathcal{B}_\delta}(\mathbf{x}_{k(i)}^\perp/r)|f|^2 \right] \prod_{k=1}^{n} b_r(\mathbf{x}_k^\perp)^2 d\mathbf{x}_k^\perp . \tag{8.54}$$

To estimate this term from below, we use Temple's inequality, as in Sect. 2.2. Let $\tilde{e}^\perp$ denote the gap above zero in the spectrum of $-\Delta^\perp + V^\perp - e^\perp$, i.e., the lowest non-zero eigenvalue. By scaling, $\tilde{e}^\perp/r^2$ is the gap in the spectrum of $-\Delta^\perp + V_r^\perp - e^\perp/r^2$. Note that under the transformation $\phi \mapsto b_r^{-1}\phi$ this latter operator is unitarily equivalent to $\nabla^{\perp*} \cdot \nabla^\perp$ as an operator on $L^2(\mathbb{R}^2, b_r(\mathbf{x}^\perp)^2 d\mathbf{x}^\perp)$, as considered in (8.54). Hence also this operator has $\tilde{e}^\perp/r^2$ as its energy gap. Denoting

$$\langle U^k \rangle = \int \left(\sum_{i=1}^{n} U(|\mathbf{x}_i - \mathbf{x}_{k(i)}|)\chi_{\mathcal{B}_\delta}(\mathbf{x}_{k(i)}^\perp/r) \right)^k \prod_{k=1}^{n} b_r(\mathbf{x}_k^\perp)^2 d\mathbf{x}_k^\perp , \tag{8.55}$$

Temple's inequality implies

$$(8.54) \geq |F|^2 a'\langle U \rangle \left(1 - a' \frac{\langle U^2 \rangle}{\langle U \rangle} \frac{1}{\varepsilon \tilde{e}^\perp/r^2 - a'\langle U \rangle} \right) . \tag{8.56}$$

Now, using (2.43) and Schwarz's inequality, $\langle U^2 \rangle \leq 3n(R^3 - R_0^3)^{-1}\langle U \rangle$, and

$$\langle U \rangle \leq n(n - 1)\frac{\|b\|_4^4}{r^2} \frac{3\pi R^2}{R^3 - R_0^3} . \tag{8.57}$$

Therefore

$$(8.56) \geq |F|^2 a''\langle U \rangle \,, \tag{8.58}$$

where we put all the error terms into the modified coupling constant a''. It remains to derive a lower bound on $\langle U \rangle$. Let

$$d(z - z') = \int_{\mathbb{R}^4} b_r(\mathbf{x}^\perp)^2 b_r(\mathbf{y}^\perp)^2 U(|\mathbf{x} - \mathbf{y}|) \chi_{\mathcal{B}_\delta}(\mathbf{y}^\perp / r) d\mathbf{x}^\perp d\mathbf{y}^\perp \,. \tag{8.59}$$

Note that $d(z) = 0$ if $|z| \geq R$. An estimate similar to (2.49) gives

$$\langle U \rangle \geq \sum_{i \neq j} d(z_i - z_j) \left(1 - (n-2) \frac{\pi R^2}{r^2} \|b\|_\infty^2 \right) \,. \tag{8.60}$$

Note that, for an appropriate choice of R, d is close to a δ-function with the desired coefficient. To make the connection with the δ-function, we can use a bit of the kinetic energy saved in (8.53) to obtain

$$\int \left[\frac{\varepsilon}{n-1} |\partial_i F|^2 + a''' d(z_i - z_j) |F|^2 \right] dz_i$$

$$\geq \tfrac{1}{2} g' \max_{|z_i - z_j| \leq R} |F|^2 \chi_{[R, \ell - R]}(z_j) \left(1 - \left(\frac{2(n-1)}{\varepsilon} g' R \right)^{1/2} \right) \,. \tag{8.61}$$

Putting all the previous estimates together, we arrive at

$$\langle \Psi | H | \Psi \rangle - \frac{n e^\perp}{r^2} \geq \sum_{i=1}^n \int \left[(1 - \varepsilon) |\partial_i F|^2 \chi_{\min_k |z_i - z_k| \geq R}(z_i) \right] \prod_{k=1}^n dz_k$$

$$+ \sum_{i \neq j} \tfrac{1}{2} g'' \int \max_{|z_i - z_j| \leq R} |F|^2 \chi_{[R, \ell - R]}(z_j) \prod_{k, \, k \neq i} dz_k \tag{8.62}$$

for an appropriate coupling constant g'' that contains all the error terms. Now assume that $(n+1)R < \ell$. Given an F with $\int |F|^2 dz_1 \cdots dz_n = 1$, define, for $0 \leq z_1 \leq z_2 \leq \cdots \leq z_n \leq \ell - (n+1)R$,

$$\psi(z_1, \dots, z_n) = F(z_1 + R, z_2 + 2R, z_3 + 3R, \dots, z_n + nR) \,, \tag{8.63}$$

and extend the function to all of $[0, \ell - (n+1)R]^n$ by symmetry. A simple calculation shows that

$$(8.62) \geq \langle \psi | H' | \psi \rangle \geq (1 - \varepsilon) E_N^{1D}(n, \ell - (n+1)R, g'') \langle \psi | \psi \rangle$$

$$\geq (1 - \varepsilon) E_N^{1D}(n, \ell, g'') \langle \psi | \psi \rangle \,, \tag{8.64}$$

where H' is the Hamiltonian (8.3) with a factor $(1 - \varepsilon)$ in front of the kinetic energy term.

It remains to estimate $\langle \psi | \psi \rangle$. Using that F is related to the true ground state Ψ by (8.52), we can estimate it in terms of the total QM energy, namely

$$
\langle \psi | \psi \rangle \;\geq\; 1 - \frac{2R}{g''} \left(E_N^{QM}(N, \ell, r, a) - \frac{ne^\perp}{r^2} \right)
$$
$$
-2n\frac{R}{\ell} - 4nR \left(\frac{1}{n} E_N^{QM}(n, \ell, r, a) - \frac{e^\perp}{r^2} \right)^{1/2}. \qquad (8.65)
$$

Collecting all the error terms and choosing

$$
R = r \left(\frac{a}{r} \right)^{1/4} \;,\qquad \varepsilon = \left(\frac{a}{r} \right)^{1/8} \;,\qquad \delta = \left(\frac{a}{r} \right)^{1/8} \;, \qquad (8.66)
$$

(8.64) and (8.65) lead to the desired lower bound.   □

As already noted above, Lemma 8.4 is the key to the proof of Theorems 8.1 and 8.2. The estimates are used in each box, and the particles are distributed optimally among the boxes. For the global lower bound, superadditivity of the energy and convexity of the energy density $\rho^3 e(g/\rho)$ are used, generalizing corresponding arguments in Chapter 2. We refer to [LSeY6] for details.

Chapter 9

Two-Dimensional Behavior in Disc-Shaped Traps

In this chapter, which is based on [SY], we discuss the dimensional reduction of a Bose gas in a trap that confines the particles strongly in one direction so that two-dimensional behavior is expected. There are many similarities with the emergence of one-dimensional behavior in cigar-shaped traps discussed in the previous chapter but also some notable differences. As in the case of cigar-shaped traps, there is a basic division of the parameter domain into two regions (this is also noted in [PHS] and [PiSt2]): one where a limit of a three-dimensional Gross-Pitaevskii theory applies, and a complementary region described by a "truly" low dimensional theory. In the case discussed in Chapter 8 the latter is a density functional theory based on the exact Lieb-Liniger solution for the energy of a strongly interacting (and highly correlated) one-dimensional gas with delta interactions. (Note that in 1D strong interactions means low density.) In the present case, on the other hand, the gas is *weakly* interacting in all parameter regions. In the region not accessible from 3D GP theory the energy formula (3.1) for a dilute two-dimensional Bose gas with a logarithmic dependence on the density applies. To enter this region extreme dilution is required. The Lieb-Liniger region in the 1D case demands also quite dilute systems, but the requirement is even more stringent in 2D. This will be explained further below.

We recall from Chapter 3 that the energy per particle of a dilute, homogeneous, two-dimensional Bose gas with density ρ_{2D} and scattering length a_{2D} of the interaction potential is (in units such that $\hbar = 2m = 1$)

$$e_{2D} \approx 4\pi \rho_{2D} |\ln(\rho_{2D} a_{2D}^2)|^{-1}. \tag{9.1}$$

The corresponding result in three dimensions is

$$e_{3D} \approx 4\pi \rho_{3D} a_{3D} \tag{9.2}$$

as discussed in Chapter 2. In the following we shall denote the two-dimensional density, ρ_{2D}, simply by ρ and the three dimensional scattering length, a_{3D}, by a. The basic message of this chapter is that when the thickness, h, of the trap tends to zero then Eq. (9.1) can be used with an effective two-dimensional scattering length $a_{2D} = h \exp(-(\text{const.})h/a)$. A more precise formula is given in Eq. (9.14) below. If $|\ln \rho h^2| \ll h/a$, then $|\ln(\rho a_{2D}^2)| \approx h/a$, and the two-dimensional formula (9.1) leads to the same result as the three dimensional formula (9.2), because $\rho_{3D} \sim \rho/h$. The "true" two dimensional region requires $|\ln \rho h^2| \gtrsim h/a$ and hence the condition $\rho^{-1/2} \gtrsim h e^{h/a}$ for the interparticle distance, $\rho^{-1/2}$. This should be compared with the corresponding condition for the 1D Lieb-Liniger region of the previous chapter where the interparticle distance is "only" required to be of the order or larger than h^2/a.

The formula $a_{2D} = h \exp(-(\text{const.})h/a)$ for the scattering length appeared first in [PHS]. It can be motivated by considering a weak, bounded potential, where perturbation theory can be used to compute the energy (C.8) in Appendix C that is directly related to the scattering length. This perturbative calculation is carried out in Section 9.3 as a step in the proof of a lower bound for the many-body energy; its relation to the formula for a_{2D} is explained in the remark after Corollary 9.4.

We now define the setting and state the results more precisely. We consider N identical, spinless bosons in a confining, three-dimensional trap potential and with a repulsive, rotationally symmetric pair interaction. We take the direction of strong confinement as the z-direction and write the points $\mathbf{x} \in \mathbb{R}^3$ as (x, z), $x \in \mathbb{R}^2$, $z \in \mathbb{R}$. The Hamiltonian is

$$H_{N,L,h,a} = \sum_{i=1}^{N} (-\Delta_i + V_{L,h}(\mathbf{x}_i)) + \sum_{1 \le i < j \le N} v_a(|\mathbf{x}_i - \mathbf{x}_j|) \tag{9.3}$$

with

$$V_{L,h}(\mathbf{x}) = V_L(x) + V_h^{\perp}(z) = \frac{1}{L^2} V(L^{-1}x) + \frac{1}{h^2} V^{\perp}(h^{-1}z), \tag{9.4}$$

$$v_a(|\mathbf{x}|) = \frac{1}{a^2} v(a^{-1}|\mathbf{x}|). \tag{9.5}$$

The confining potentials V and $V^\perp$ are assumed to be locally bounded and tend to ∞ as $|x|$ and $|z|$ tend to ∞. The interaction potential v is assumed to be nonnegative, of finite range and with scattering length 1; the scaled potential v_a then has scattering length a. We regard v, $V^\perp$ and V as fixed and L, h, a as scaling parameters. The Hamiltonian (9.3) acts on symmetric wave functions in $L^2(\mathbb{R}^{3N}, d\mathbf{x}_1 \cdots d\mathbf{x}_N)$. Its ground state energy, $E^{\mathrm{QM}}(N, L, h, a)$, scales with L as

$$E^{\mathrm{QM}}(N, L, h, a) = \frac{1}{L^2} E^{\mathrm{QM}}(N, 1, h/L, a/L). \tag{9.6}$$

Taking $N \to \infty$ but keeping h/L and Na/L fixed leads to a three dimensional Gross-Pitaevskii description of the ground state as proved in Chapter 6. The corresponding energy functional is (cf. (6.2))

$$\mathcal{E}_{\mathrm{3D}}^{\mathrm{GP}}[\phi] = \int_{\mathbb{R}^3} \left\{ |\nabla\phi(\mathbf{x})|^2 + V_{L,h}(\mathbf{x})|\phi(\mathbf{x})|^2 + 4\pi Na|\phi(\mathbf{x})|^4 \right\} d^3\mathbf{x} \tag{9.7}$$

and the energy per particle is

$$
\begin{aligned}
E_{\mathrm{3D}}^{\mathrm{GP}}(N, L, h, a)/N &= \inf\{\mathcal{E}_{\mathrm{3D}}^{\mathrm{GP}}[\phi], \textstyle\int |\phi(\mathbf{x})|^2 d^2\mathbf{x} = 1\} \\
&= (1/L^2)E_{\mathrm{3D}}^{\mathrm{GP}}(1, 1, h/L, Na/L).
\end{aligned}
\tag{9.8}
$$

By the GP limit theorem, Thm. 6.1, we have, for fixed h/L and Na/L,

$$\lim_{N\to\infty} \frac{E^{\mathrm{QM}}(N, L, h, a)}{E_{\mathrm{3D}}^{\mathrm{GP}}(N, L, h, a)} = 1. \tag{9.9}$$

It is important to note, however, that the estimates in Chapter 6 are not uniform in the ratio h/L and the question what happens if $h/L \to 0$ is not addressed in Ch. 6. It will be shown in the next section that a *part* of the parameter range for thin traps can be treated by considering, at fixed Na/h, the $h/L \to 0$ limit of $E_{\mathrm{3D}}^{\mathrm{GP}}(1, 1, h/L, Na/L)$ with the ground state energy for the transverse motion, $\sim 1/h^2$, subtracted. But this limit can evidently never lead to a logarithmic dependence on the density and it does not give the correct limit formula for the energy in the whole parameter range.

To cover all cases we have to consider a two-dimensional Gross-Pitaevskii theory of the type studied in Section 6.2, i.e.,

$$\mathcal{E}_{\mathrm{2D}}^{\mathrm{GP}}[\varphi] = \int_{\mathbb{R}^2} \left\{ |\nabla\varphi(x)|^2 + V_L(x)|\varphi(x)|^2 + 4\pi Ng|\varphi(x)|^4 \right\} d^2x \tag{9.10}$$

with

$$g = |\ln(\bar\rho a_{\mathrm{2D}}^2)|^{-1}. \tag{9.11}$$

Here $\bar{\rho}$ is the mean density, defined by Eq. (6.70). For simplicity we shall assume that V is homogeneous of some order $s > 0$, i.e., $V(\lambda x) = \lambda^s V(x)$, and in this case

$$\bar{\rho} \sim N^{s/(s+2)}/L^2 = N/\bar{L}^2 \quad \text{with} \quad \bar{L} = N^{1/(s+2)}L. \tag{9.12}$$

The length $\bar{L}$ measures the effective extension of the gas cloud in the two-dimensional trap. A box potential corresponds to $L = \bar{L}$, i.e., $s = \infty$ and hence $\bar{\rho} \sim N/L^2$. The energy per particle corresponding to (9.10) is

$$E_{\text{2D}}^{\text{GP}}(N,L,g)/N = \inf\{\mathcal{E}_{\text{2D}}^{\text{GP}}[\varphi], \int |\varphi(x)|^2 d^2x = 1\} = (1/L^2)E_{\text{2D}}^{\text{GP}}(1,1,Ng). \tag{9.13}$$

Let s_h be the normalized ground state wave function of the one-particle Hamiltonian $-d^2/dz^2 + V_h^\perp(z)$. It can be written as $s_h(z) = h^{-1/2}s(h^{-1}z)$ and the ground state energy as $e_h^\perp = h^{-2}e^\perp$, where $s(z)$ and $e^\perp$ are, respectively, the ground state wave function and ground state energy of $-d^2/dz^2 + V^\perp(z)$. We *define* the effective two dimensional scattering length by the formula

$$a_{\text{2D}} = h \exp\left(-(\int s(z)^4 dz)^{-1}h/2a\right). \tag{9.14}$$

Then, using (9.11),

$$g = |-\ln(\bar{\rho}h^2) + (\int s(z)^4 dz)^{-1}h/a|^{-1}. \tag{9.15}$$

The justification of the definition (9.14) is Theorem 9.1 below.

Remark. Since a_{2D} appears only under a logarithm, and $a/h \to 0$, one could, at least as far as leading order computations are concerned, equally well define the two dimensional scattering length as $a_{\text{2D}}' = b \exp\left(-(\int s(z)^4 dz)^{-1}h/2a\right)$ with b satisfying $c\, a \le b \le C h$ for some constants $c > 0$, $C < \infty$. In fact, if $g' = |\ln(\bar{\rho}(a_{\text{2D}}')^2)|^{-1}$, then

$$\frac{g}{g'} = 1 + \frac{2\ln(b/h)}{|-\ln(\bar{\rho}h^2) + (\text{const.})h/a|} \to 1 \tag{9.16}$$

because $(a/h)\ln(b/h) \to 0$.

We can now state the main result of this chapter:

Theorem 9.1 (From 3D to 2D). *Let $N \to \infty$ and at the same time $h/L \to 0$ in such a way that $h^2\bar{\rho}g \to 0$ (with g given by Eq. (9.15)). Then*

$$\lim \frac{E^{\text{QM}}(N,L,h,a) - Nh^{-2}e^\perp}{E_{\text{2D}}^{\text{GP}}(N,L,g)} = 1. \tag{9.17}$$

Remarks: 1. The condition $h^2 \bar{\rho} g \to 0$ means that the ground state energy $h^{-2} e^{\perp}$ associated with the confining potential in the z-direction is much larger than the energy $\bar{\rho} g$. This is the condition of *strong confinement* in the z-direction. In the case that $h/a \gg |\ln(\bar{\rho} h^2)|$ we have $g \sim a/h$ and hence the condition in that region is equivalent to

$$\bar{\rho} a h \ll 1. \tag{9.18}$$

On the other hand, if $h/a \lesssim |\ln(\bar{\rho} h^2)|$ the strong confinement condition is equivalent to $h^2 \bar{\rho} |\ln(h^2 \bar{\rho})|^{-1} \ll 1$, which means simply that

$$\bar{\rho} h^2 \ll 1. \tag{9.19}$$

Both (9.18) and (9.19) clearly imply $\bar{\rho} a_{2D}^2 \ll 1$, i.e., the gas is dilute in the 2D sense (and also in the 3D sense, $\rho_{3D} a^3 \ll 1$, because $\rho_{3D} = \rho/h$). This is different from the situation in cigar-shaped traps considered in Chapter 8 where the gas can be either dilute or dense in the 1D sense, depending on the parameters (although it is always dilute in the 3D sense).

2. It is, in fact, not necessary to demand $h/L \to 0$ explicitly in Theorem 9.1. The reason is as follows. In the region where $h/a \lesssim |\ln(\bar{\rho} h^2)|$, the strong confinement condition $\bar{\rho} h^2 \ll 1$ immediately implies $h/L \ll 1$ because $\bar{\rho} \gg 1/L^2$, cf. Eq. (9.12). If $h/a \gg |\ln(\bar{\rho} h^2)|$, then at least $\bar{\rho} a h \ll 1$ holds true. This leaves only the alternatives $h/L \to 0$, or, if h/L stays bounded away from zero, $Na/L \to 0$. But the latter alternative means, by the three dimensional Gross-Pitaevskii limit theorem, that the energy converges to the energy of a noninteracting, trapped gas, for which (9.17) obviously holds true.

We shall refer to the parameter region where $h/a \gg |\ln(\bar{\rho} h^2)|$ as **Region I**, and the one where $h/a \lesssim |\ln(\bar{\rho} h^2)|$ as **Region II**. In Region I we can take

$$g = (\int s(z)^4 dz) a/h. \tag{9.20}$$

In Region II $g \sim |\ln(\bar{\rho} h^2)|^{-1}$, and in the extreme case that $h/a \ll |\ln(\bar{\rho} h^2)|$,

$$g = |\ln(\bar{\rho} h^2)|^{-1}. \tag{9.21}$$

In particular g is then independent of a (but dependent on $\bar{\rho}$). As remarked earlier, Region II is only relevant for very dilute gases since it requires interparticle distances $\bar{\rho}^{-1/2} \gtrsim h e^{h/a}$.

By Eq. (9.13) the relevant coupling parameter is Ng rather than g itself, and both Region I and Region II can be divided further, according to $Ng \ll 1$,

$Ng \sim 1$, or $Ng \gg 1$. The case $Ng \ll 1$ corresponds simply to an ideal gas in the external trap potential. Note that this limit can both be reached from Region I by taking $a/h \to 0$ at fixed $\bar{\rho}h^2$, or from Region II by letting $\bar{\rho}h^2$ tend more rapidly to zero than $e^{-h/a}$. The case $Ng \sim 1$ in Region I corresponds to a GP theory with coupling parameter $\sim Na/h$ as was already explained, in particular after Eq. (9.9). The case $Ng \gg 1$ is the 'Thomas-Fermi' case where the gradient term in the energy functional (9.10) can be ignored. In Region II, the case $Ng \lesssim 1$ requires $\bar{\rho}^{-1/2} \gtrsim he^N$ and is thus only of academic interest, while $\bar{\rho}^{-1/2} \ll he^N$ (but still $he^{h/a} \lesssim \bar{\rho}^{-1/2}$) corresponds to the TF case.

The subdivision of the parameter range just described is somewhat different from the situation described in Chapter 8. This is due to the different form of the energy per particle of the low dimensional gas as function of the density.

9.1   The 2D Limit of 3D GP Theory

As in Section 8.2 certain aspects of the dimensional reduction of the many-body system can be seen already in the much simpler context of GP theory. In this section we consider the $h/L \to 0$ limit of the 3D GP ground state energy. The result is, apart from the confining energy, the 2D GP energy with coupling constant $g \sim a/h$. This shows in particular that Region II, where $g \sim |\ln(\bar{\rho}h^2)|^{-1}$, cannot be reached as a limit of 3D GP theory.

Theorem 9.2 (2D limit of 3D GP energy). *Define $g = \left(\int s(z)^4 dz \right) a/h$. If $h/L \to 0$, then*

$$\frac{E_{3D}^{GP}(N, L, h, a) - Nh^{-2}e^{\perp}}{E_{2D}^{GP}(N, L, g)} \to 1 \qquad (9.22)$$

uniformly in the parameters, as long as $\bar{\rho}ah \to 0$.

Remark. Since $E_{2D}^{GP}(1, L, Ng) \sim L^{-2} + \bar{\rho}a/h$, the condition $\bar{\rho}ah \to 0$ is equivalent to $h^2 E_{2D}^{GP}(1, L, Ng) \to 0$, which means simply that the 2D GP energy per particle is much less than the confining energy, $\sim 1/h^2$.

Proof. The proof is analogous to that of Thm. 8.3. Because of the scaling relation (9.8) it suffices to consider the case $N = 1$ and $L = 1$.

For an upper bound to the 3D GP ground state energy we make the ansatz

$$\phi(\mathbf{x}) = \varphi_{GP}(x)s_h(z), \qquad (9.23)$$

where φ_{GP} is the minimizer of the 2D GP functional with coupling constant g. Then

$$\mathcal{E}_{3D}^{GP}[\phi] = e^{\perp}/h^2 + E_{2D}^{GP}(1,1,g) \tag{9.24}$$

and hence

$$E_{3D}^{GP}(1,1,h,a) - e^{\perp}/h^2 \le E_{2D}^{GP}(1,1,g). \tag{9.25}$$

For the lower bound we consider the one-particle Hamiltonian (in 3D)

$$H_{h,a} = -\Delta + V_{1,h}(\mathbf{x}) + 8\pi a|\varphi_{GP}(x)|^2 s_h(z)^2. \tag{9.26}$$

Taking the 3D GP minimizer Φ as a test state gives

$$\begin{aligned}
\inf \operatorname{spec} H_{h,a} \quad &\le \quad E_{3D}^{GP}(1,1,h,a) - 4\pi a \int_{\mathbb{R}^3} |\Phi(\mathbf{x})|^4 d^3\mathbf{x} \\
&\quad + 8\pi a \int_{\mathbb{R}^3} |\varphi_{GP}(x)|^2 s_h(z)^2 |\Phi(\mathbf{x})|^2 d^3\mathbf{x} \\
&\le \quad E_{3D}^{GP}(1,1,h,a) + 4\pi a \int_{\mathbb{R}^3} |\varphi_{GP}(x)|^4 s_h(z)^4 d^3\mathbf{x} \\
&= \quad E_{3D}^{GP}(1,1,h,a) + 4\pi g \int_{\mathbb{R}^2} |\varphi_{GP}(x)|^4 d^2 x. \tag{9.27}
\end{aligned}$$

To bound $H_{h,a}$ from below we consider first for fixed $x \in \mathbb{R}^2$ the Hamiltonian (in 1D)

$$H_{h,a,x} = -\partial_z^2 + V_h^{\perp}(z) + 8\pi a|\varphi_{GP}(x)|^2 s_h(z)^2. \tag{9.28}$$

We regard $-\partial_z^2 + V_h^{\perp}(z)$ as its "free" part and $8\pi a|\varphi_{GP}(x)|^2 s_h(z)^2$ as a perturbation. Since the perturbation is positive all eigenvalues of $H_{h,a,x}$ are at least as large as those of $-\partial_z^2 + V_h(z)$; in particular the first excited eigenvalue is $\sim 1/h^2$. The expectation value in the ground state s_h of the free part is

$$\langle H_{h,a,x}\rangle = e^{\perp}/h^2 + 8\pi g|\varphi_{GP}(x)|^2. \tag{9.29}$$

Temple's inequality (2.51) gives

$$H_{h,a,x} \ge \left(e^{\perp}/h^2 + 8\pi g|\varphi_{GP}(x)|^2\right)\left(1 - \frac{\langle(H_{h,a,x} - \langle H_{h,a,x}\rangle)^2\rangle}{\langle H_{h,a,x}\rangle(\tilde{e}^{\perp} - e^{\perp})/h^2}\right), \tag{9.30}$$

where $\tilde{e}^{\perp}/h^2$ is the lowest eigenvalue above the ground state energy of $-\partial_z^2 + V_h^{\perp}(z)$. Since

$$H_{h,a,x}s_h = (e^{\perp}/h^2)s_h + 8\pi a|\varphi_{GP}(x)|^2 s_h^3 \tag{9.31}$$

we have $(H_{h,a,x} - \langle H_{h,a,x} \rangle)s_h = 8\pi|\varphi_{GP}(x)|^2(as_h^3 - gs_h)$ and hence, using $g = a \int s_h^4 = (a/h) \int s^4$,

$$
\begin{aligned}
\langle ((H_{h,a,x} - \langle H_{h,a,x} \rangle)^2) \rangle &= (8\pi)^2|\varphi_{GP}(x)|^4 \int \left(as_h(z)^3 - gs_h(z) \right)^2 dz \\
&\leq (8\pi)^2\|\varphi_{GP}\|_\infty^4 (a/h)^2 \left[\int s^6 - \left(\int s^4 \right)^2 \right] \\
&\leq \text{const.} \, E_{2D}^{GP}(1,1,g)^2
\end{aligned}
\tag{9.32}
$$

where we have used Lemma 2.1 in [LSeY2] to bound the term $g\|\varphi_{GP}\|_\infty^2$ by const. $E_{2D}^{GP}(1,1,g)$. We thus see from (9.29) and the assumption $h^2 E_{2D}^{GP}(1,1,g) \to 0$ that the error term in the Temple inequality (9.30) is $o(1)$.

Now $H_{h,a} = -\Delta_x + V(x) + H_{h,a,x}$, so from (9.30) we conclude that

$$
H_{h,a} \geq \left((e^\perp/h^2) - \Delta_x + V(x) + 8\pi g|\varphi_{GP}(x)|^2 \right)(1 - o(1)).
\tag{9.33}
$$

On the other hand, the lowest energy of $-\Delta_x + V(x) + 8\pi g|\varphi_{GP}(x)|^2$ is just $E_{2D}^{GP}(1,1,g) + 4\pi g \int_{\mathbb{R}^2} |\varphi_{GP}(x)|^4 d^2x$. Combining (9.27) and (9.33) we thus get

$$
E_{3D}^{GP}(1,1,h,a) - e^\perp/h^2 \geq E_{2D}^{GP}(1,1,g)(1 - o(1)).
\tag{9.34}
$$

$\square$

Remark. This proof holds also for the Gross-Pitaevskii functional for rotating gases, i.e., if a rotational term, $-\langle \phi, \vec{\Omega} \cdot \vec{L}\phi \rangle$ is added to the functional. Here $\vec{\Omega}$ is the angular velocity, assumed to point in the z-direction, and $\vec{L}$ the angular momentum operator. The minimizer φ_{GP} is in this case complex valued in general and may not be unique [Se4].

9.2   Upper Bound

We now turn to the many-body problem, i.e., the proof of Theorem 9.1. For simplicity we shall here only discuss the situation where the system is homogeneous in the 2D variables x, i.e., where the confining potential $V_L(x)$ is replaced by a large box whose side length L is taken to infinity in a thermodynamic limit. An inhomogeneous system in the x directions can be treated by analogous methods if one in a first step separates out a factor $\prod_i \varphi_{GP}(x_i)$ in the wave function with φ_{GP} the minimizer of the 2D GP functional. (This is the same technique as used in Chapter 6.)

As in the problem discussed in Chapter 8 the key lemmas are energy bounds in boxes with *finite* particle number. The bounds for the total system are obtained by distributing the particles optimally among the boxes. We shall here focus on the estimates for the individual boxes, starting with the upper bound.

Consider the Hamiltonian

$$H = \sum_{i=1}^{n} \left(-\Delta_i + V_h^{\perp}(z_i) \right) + \sum_{1 \le i < j \le n} v_a(|\mathbf{x}_i - \mathbf{x}_j|) \tag{9.35}$$

in a region $\Lambda = \Lambda_2 \times \mathbb{R}$ where Λ_2 denotes a box of side length ℓ in the 2D x variables. For the upper bound on the ground state energy of (9.35) we impose *Dirichlet* boundary conditions on the 2D Laplacian. The goal is to prove, for a given 2D density ρ and parameters a and h, that for a suitable choice of ℓ the energy per particle is bounded above by

$$4\pi\rho |\ln(\rho a_{2D}^2)|^{-1}(1 + o(1)) \tag{9.36}$$

where $\rho = n/\ell^2$ and a_{2D} is given by Eq. (9.14). Moreover, the Dirichlet localization energy per particle, $\sim 1/\ell^2$, should be small compared to (9.36). The relative error, $o(1)$, in (9.36) tends to zero with the small parameters a/h and ρah (Region I), or a/h and ρh^2 (Region II).

The choice of variational functions depends on the parameter regions and we are first concerned with the Region II, i.e., the case $|\ln(\rho h^2)| \gtrsim h/a$.

Let $f_0(r)$ be the solution of the zero energy scattering equation

$$-\Delta f_0 + \tfrac{1}{2} v_a f_0 = 0, \tag{9.37}$$

normalized so that $f_0(r) = (1 - a/r)$ for $r \ge R_0$. It satisfies $0 \le f_0(r) \le 1$ and $0 \le f_0'(r) \le \min\{1/r, a/r^2\}$. For $R > R_0$ we define $f(r) = f_0(r)/(1 - a/R)$ for $0 \le r \le R$, and $f(r) = 1$ for $r > R$. Define a two-dimensional potential by

$$W(x) = \frac{2\|s\|_4^4}{h} \int_{\mathbb{R}} \left[f'(|\mathbf{x}|)^2 + \tfrac{1}{2} v_a(|\mathbf{x}|) f(|\mathbf{x}|)^2 \right] dz. \tag{9.38}$$

Clearly, $W(x) \ge 0$, and W is rotationally symmetric with $W(x) = 0$ for $|x| \ge R$. Moreover, by partial integration, using (9.37), it follows that $W \in L^1(\mathbb{R}^2)$ with

$$\int_{\mathbb{R}^2} W(x)dx = \frac{8\pi a \|s\|_4^4}{h}(1 - a/R)^{-1}. \tag{9.39}$$

Define, for $b > R$,

$$\varphi(r) = \begin{cases} \ln(R/a_{2D})/\ln(b/a_{2D}) & \text{if} \quad 0 \le r \le R \\ \ln(r/a_{2D})/\ln(b/a_{2D}) & \text{if} \quad R \le r \le b \\ 1 & \text{if} \quad b \le r \end{cases} \tag{9.40}$$

As test function for the three dimensional Hamiltonian (9.35) we shall take

$$\Psi(\mathbf{x}_1,\ldots,\mathbf{x}_n) = F(\mathbf{x}_1,\ldots,\mathbf{x}_n)G(\mathbf{x}_1,\ldots,\mathbf{x}_n) \qquad (9.41)$$

with

$$F(\mathbf{x}_1,\ldots,\mathbf{x}_n) = \prod_{i<j} f(|\mathbf{x}_i-\mathbf{x}_j|) \quad \text{and} \quad G(\mathbf{x}_1,\ldots,\mathbf{x}_n) = \prod_{i<j} \varphi(|x_i-x_j|) \prod_{k=1}^{n} s_h(z_k).$$
$$(9.42)$$

The parameters R, b and also ℓ will eventually be chosen so that the errors compared to the expected leading term in the energy are small.

As it stands, the function (9.41) does not satisfy Dirichlet boundary conditions but this can be taken care of by multiplying the function with additional factors at energy cost $\sim 1/\ell^2$ per particle, that will turn out to be small compared to the energy of (9.41).

Since $f(|\mathbf{x}_i - \mathbf{x}_j|)\varphi(|x_i - x_j|) = 1$ for $|x_i - x_j| \geq b$ and s_h is normalized, the norm of Ψ can be estimated as (cf. the analogous Eq. (8.34))

$$\langle\Psi|\Psi\rangle \geq \ell^{2n}\left[1 - \frac{\pi n(n-1)}{2}\frac{b^2}{\ell^2}\right]. \qquad (9.43)$$

Next we consider the expectation value of H with the wave function Ψ. By partial integration we have, for every j,

$$\int |\nabla_j(FG)|^2$$

$$= \int G^2|\nabla_j F|^2 - \int F^2 G\Delta_j G = \int G^2|\nabla_j F|^2 - \int F^2 G\,\partial_{z_j}^2 G - \int F^2 G(\Delta_j^\| G)$$

$$= \int G^2|\nabla_j F|^2 - \int F^2 G\,\partial_{z_j}^2 G + \int F^2|\nabla_j^\| G|^2 + 2\int FG(\nabla_j^\| F)\cdot(\nabla_j^\| G) \quad (9.44)$$

where $\Delta_j^\|$ and $\nabla_j^\|$ are, respectively, the two dimensional Laplace operator and gradient. The term $-\int F^2 G\,\partial_{z_j}^2 G$ together with $\int V_h^\perp F^2 G^2$ gives the confinement energy, $(e^\perp/h^2)\|\Psi\|^2$.

Next we consider the first and the third term in (9.44). Since $0 \leq f \leq 1$, $f' \geq 0$ and s_h is normalized, we have

$$\sum_j \int F^2|\nabla_j^\| G|^2 + \sum_j \int |\nabla_j F|^2 G \leq \sum_j \int |\nabla_j^\| \Phi|^2 + 2\sum_{i<j} f'(|\mathbf{x}_i - \mathbf{x}_j|)^2 G^2$$

$$+ 4\sum_{k<i<j} \int f'(|\mathbf{x}_k - \mathbf{x}_i|)f'(|\mathbf{x}_k - \mathbf{x}_j|)G^2$$

$$(9.45)$$

where we have denoted $\prod_{i<j} \varphi(|x_i - x_j|)$ by Φ for short.

By Young's inequality

$$2 \int_{\mathbb{R}^2} f'(|\mathbf{x}_i - \mathbf{x}_j|)^2 s_h(z_i)^2 s_j(z_j)^2 dz_i dz_j \leq \frac{2\|s\|_4^4}{h} \int_{\mathbb{R}} f'(|(x_i - x_j, z)|)^2 dz. \quad (9.46)$$

The right side gives rise to the first of the two terms in the formula (9.38) for the two dimensional potential W. The other part is provided by $\int F^2 G^2 v_a(\mathbf{x}_i - \mathbf{x}_j)$, again using that $0 \leq f \leq 1$, $0 \leq \varphi \leq 1$ and Young's inequality.

Altogether we obtain

$$\langle \Psi | H | \Psi \rangle - (n e^{\perp}/h^2) \langle \Psi | \Psi \rangle \leq \sum_j \int_{\Lambda_2^n} |\nabla_j^{\|} \Phi|^2 + \sum_{i<j} \int_{\Lambda_2^n} W(x_i - x_j) \Phi^2 + \mathcal{R}_1 + \mathcal{R}_2$$

$$(9.47)$$

with

$$\mathcal{R}_1 = 2 \int_{\Lambda^n} FG(\nabla_j^{\|} F) \cdot (\nabla_j^{\|} G) \quad (9.48)$$

and

$$\mathcal{R}_2 = 4 \sum_{k<i<j} \int_{\Lambda^n} f'(|\mathbf{x}_k - \mathbf{x}_i|) f'(|\mathbf{x}_k - \mathbf{x}_j|) G^2 \quad (9.49)$$

$$\leq \frac{2}{3} n(n-1)(n-2) \ell^{2(n-3)} \int_{\Lambda^3} f'(|\mathbf{x}_1 - \mathbf{x}_2|) f'(|\mathbf{x}_2 - \mathbf{x}_3|) s_h(z_1)^2 s_h(z_2)^2 s_h(z_3)^2.$$

The error term $\mathcal{R}_1$ is easily dealt with: It is zero because $\varphi(r)$ is constant for $r \leq R$ and $f(r)$ is constant for $r \geq R$.

The other term, $\mathcal{R}_2$, is estimated as follows. Since $f'(r) = 0$ for $r \geq R$ we can use the Cauchy Schwarz inequality for the integration over $\mathbf{x}_1$ at fixed $\mathbf{x}_2$ to obtain

$$\int f'(|\mathbf{x}_1 - \mathbf{x}_2|) s_h(z_1)^2 d\mathbf{x}_1$$

$$\leq \left(\int f'(|\mathbf{x}_1 - \mathbf{x}_2|)^2 d\mathbf{x}_1 \right)^{1/2} \left(\int_{|\mathbf{x}_1 - \mathbf{x}_2| \leq R} s_h(z_1)^4 d\mathbf{x}_1 \right)^{1/2}$$

$$\leq (4\pi \|s\|_\infty a' R^3 / 3h^2)^{1/2} \quad (9.50)$$

with $a' = a(1 - a/R)^{-1}$. The same estimate for the integration over $\mathbf{x}_3$ and a subsequent integration over $\mathbf{x}_2$ gives

$$\mathcal{R}_2 \leq (\text{const.}) \ell^{2n} n^3 \frac{a' R^3}{\ell^4 h^2}. \quad (9.51)$$

We need $\mathcal{R}_2/\langle\Psi|\Psi\rangle$ to be small compared to the leading term in the energy, $\sim n^2\ell^{-2}|\ln(\rho h^2)|^{-1}$ with $\rho = n/\ell^2$. (Recall that we are in Region II where $|\ln(\rho h^2)| \gtrsim h/a$.) Moreover, the leading term should be large compared to the Dirichlet localization energy, which is $\sim n/\ell^2$. We are thus lead to the conditions (the first comes from (9.43)):

$$\frac{n^2 b^2}{\ell^2} \ll 1, \quad \frac{na'R^3|\ln(\rho h^2)|}{\ell^2 h^2} \ll 1, \quad \frac{n}{|\ln(\rho h^2)|} \gg 1, \tag{9.52}$$

which can also be written

$$\rho^2\ell^2 b^2 \ll 1, \quad \frac{\rho a'R^3|\ln(\rho h^2)|}{h^2} \ll 1, \quad \frac{|\ln(\rho h^2)|}{\rho\ell^2} \ll 1. \tag{9.53}$$

These conditions are fulfilled if we choose

$$R = h, \quad b = \rho^{-1/2}|\ln(\rho h^2)|^{-\alpha} \tag{9.54}$$

with $\alpha > 1/2$ and

$$\rho^{-1/2}|\ln(\rho h^2)|^{1/2} \ll \ell \ll \rho^{-1/2}|\ln(\rho h^2)|^{\alpha}. \tag{9.55}$$

Note also that $n = \rho\ell^2 \gg 1$.

It remains to compare

$$\langle\Psi|\Psi\rangle^{-1}\left(\sum_j \int_{\mathbb{R}^{2n}} |\nabla_j^{\|}\Phi|^2 + \sum_{i<j} \int_{\mathbb{R}^{2n}} W(x_i - x_j)\Phi^2\right) \tag{9.56}$$

with the expected leading term of the energy, i.e., $4\pi(n^2/\ell^2)|\ln(na_{2D}^2/\ell^2)|^{-1}$.

We consider first the simplest case, i.e., $n = 2$. We have

$$\int_{\mathbb{R}^2} |\nabla^{\|}\varphi|^2 = (\ln(b/a_{2D}))^{-2}2\pi\int_R^b \frac{dr}{r} = (\ln(b/a_{2D}))^{-2}2\pi\ln(b/R), \tag{9.57}$$

$$\frac{1}{2}\int_{\mathbb{R}^2} W\varphi^2 = \frac{4\pi a\|s\|_4^4}{h}\left(\frac{\ln(R/a_{2D})}{\ln(b/a_{2D})}\right)^2. \tag{9.58}$$

Inserting the formula (9.14) for a_{2D} and using $R = h$, $b = \rho^{-1/2}|\ln(\rho h^2)|^{-\alpha}$ and $a' = a(1 + o(1))$ we have

$$\int_{\mathbb{R}^2} (|\nabla^{\|}\varphi|^2 + \tfrac{1}{2}W\varphi^2) =$$
$$2\pi(\ln(b/a_{2D}))^{-2}\left[\ln(b/h) + (h/2a'\|s\|_4^4)\right] = 4\pi|\ln(\rho/a_{2D}^2)|^{-1}(1 + o(1)). \tag{9.59}$$

For $n > 2$ we can use the symmetry of Φ to write, using (9.59) as well as $0 \leq \varphi(r) \leq 1$,

$$\sum_j \int_{\Lambda_2^n} |\nabla_j^\| \Phi|^2 + \sum_{i<j} \int_{\Lambda_2^n} W(x_i - x_j)\Phi^2$$

$$= n \left(\int_{\Lambda_2^n} |\nabla_1^\| \Phi|^2 + \tfrac{1}{2} \sum_{i=2}^n \int_{\Lambda_2^n} W(x_i - x_1)\Phi^2 \right)$$

$$\leq 4\pi n^2 \ell^{2(n-1)} |\ln(na_{2D}^2/\ell^2)|^{-1}(1 + o(1)) + \mathcal{R}_3 \tag{9.60}$$

with

$$\mathcal{R}_3 = n^3 \ell^{2(n-3)} \int_{\Lambda_2^3} \varphi'(|x_2 - x_1|)\varphi'(|x_3 - x_1|). \tag{9.61}$$

We estimate $\mathcal{R}_3$ in the same way as (9.50), obtaining

$$\mathcal{R}_3 \leq (\text{const.})\ell^{2(n-2)} n^3 b^2 (\ln(b/a_{2D}))^{-2} 2\pi \ln(b/R). \tag{9.62}$$

The condition that $\mathcal{R}_3$ has to be much smaller than the leading term, given by $4\pi n^2 \ell^{2(n-1)} |\ln(na_{2D}^2/\ell^2)|^{-1}$, is equivalent to

$$\frac{nb^2}{\ell^2} \ln(b/R) \ll 1. \tag{9.63}$$

With the choice (9.54) this holds if $\alpha > 1/2$.

In Region I the ansatz (9.41) can still be used, but this time we take $b = R$, i.e., $\varphi \equiv 1$. In this region $(a/h)|\ln(\rho h^2)| = o(1)$ and the leading term in the energy is $\sim n^2 \ell^{-2} a/h$. Conditions (9.52) are now replaced by

$$\frac{n^2 R^2}{\ell^2} \ll 1, \qquad \frac{nR^3}{\ell^2 h} \ll 1, \qquad \frac{na}{h} \gg 1 \tag{9.64}$$

where have here used that $a' = a(1 + o(1))$, provided $R \gg a$. Note that the last condition in (9.52) means in particular that $n \gg 1$. Putting again $\rho = n/\ell^2$, (9.64) can be written as

$$\rho^2 \ell^2 R^2 \ll 1, \qquad \frac{\rho R^3}{h} \ll 1, \qquad \frac{h}{\rho \ell^2 a} \ll 1. \tag{9.65}$$

By assumption, $a/h \ll 1$, but also $\rho ah \ll 1$ by the condition of strong confinement, c.f. (9.18). We take

$$R = a(\rho ah)^{-\beta} \tag{9.66}$$

with $0 < \beta$, so $R \gg a$. Further restrictions come from the conditions (9.65): The first and the last of these conditions imply together that

$$\frac{h}{a} \ll \rho\ell^2 \ll \frac{1}{\rho R^2} \tag{9.67}$$

which can be fulfilled if

$$\rho a h \ll (\rho a h)^{2\beta}. \tag{9.68}$$

i.e., if $\beta < \frac{1}{2}$. Note that this implies in particular $R \ll \rho^{-1/2}$. We can then take

$$\ell = \rho^{-1/2}(h/a)^{1/2}(\rho a h)^{-\gamma} \tag{9.69}$$

with

$$0 < \gamma < \frac{1 - 2\beta}{2}. \tag{9.70}$$

The second of the conditions (9.65) requires that

$$\frac{\rho R^3}{h} = (\rho a h)(a/h)^2(\rho a h)^{-3\beta} \ll 1, \tag{9.71}$$

which holds in any case if $\beta \leq 1/3$. A possible choice satisfying all conditions is

$$\beta = \frac{1}{3}, \quad \gamma = \frac{1}{12}. \tag{9.72}$$

The error terms (9.65) are then bounded by $(\rho a h)^{1/6}$ (first and third term) and $(a/h)^2$ (second term).

Finally, with $\Phi \equiv 1$, Eqs. (9.47), (9.43) and (9.39) give

$$\langle \Psi | \Psi \rangle^{-1} \langle \Psi | H | \Psi \rangle - (ne^{\perp}/h^2) \leq \frac{4\pi n^2}{\ell^2} \frac{a\|s\|_4^4}{h}(1 + o(1)). \tag{9.73}$$

This completes the proof of the upper bound in boxes with finite n. The upper bound for the energy per particle in the 2D thermodynamic limit is obtained by dividing $\mathbb{R}^2$ into Dirichlet boxes with side length ℓ satisfying (9.55) in Region II, or (9.67) in Region I, and distributing the particles evenly among the boxes. The choice of ℓ guarantees that the error associated with the Dirichlet localization is negligible. If the system is inhomogeneous in 2D because of a trapping potential $V_L(x)$ the distribution among the boxes has to be adjusted to the density given by the GP minimizer $\varphi_{\text{GP}}(x)$ in a similar way as in Chapter 8.

9.3 Scattering Length

As a preparation for the lower bound we consider in this section the perturbative calculation of the 2D scattering length of an integrable potential.

Consider a 2D, rotationally symmetric potential $W \geq 0$ of finite range R_0. As discussed in Appendix C the scattering length is determined by minimizing, for $R \geq R_0$, the functional

$$\mathcal{E}_R[\psi] = \int_{|x| \leq R} \{|\nabla \psi|^2 + \tfrac{1}{2} W |\psi|^2\} \tag{9.74}$$

with boundary condition $\psi = 1$ for $|x| = R$. The Euler equation (zero energy scattering equation) is

$$-\Delta \psi + \tfrac{1}{2} W \psi = 0 \tag{9.75}$$

and for $r = |x| \geq R$ the minimizer, ψ_0, is

$$\psi_0(r) = \ln(r/a_{\text{scatt}}) / \ln(R/a_{\text{scatt}}) \tag{9.76}$$

with a constant a_{scatt}. This is, by definition, the 2D scattering length for the potential W. An equivalent definition follows by computing the energy,

$$E_R = \mathcal{E}_R[\psi_0] = 2\pi / \ln(R/a_{\text{scatt}}) \tag{9.77}$$

which means that

$$a_{\text{scatt}} = R \exp(-2\pi/E_R). \tag{9.78}$$

Lemma 9.3 (Scattering length for soft potentials). *Assume $W(x) = \lambda w(x)$ with $\lambda \geq 0$, $w \geq 0$, and $w \in L^1(\mathbb{R}^2)$, with $\int w(x)d^2x = 1$. Then, for $R \geq R_0$,*

$$a_{\text{scatt}} = R \exp\left(-\frac{4\pi + \eta(\lambda, R)}{\lambda}\right) \tag{9.79}$$

with $\eta(\lambda, R) \to 0$ for $\lambda \to 0$.

Proof. The statement is, by (9.78), equivalent to

$$E_R = \tfrac{1}{2}\lambda(1 + o(1)) \tag{9.80}$$

where the error term may depend on R. The upper bound is clear by the variational principle, taking $\psi = 1$ as a test function. For the lower bound note first that $\psi_0 \leq 1$. This follows from the variational principle: Since $W \geq 0$ the function $\tilde{\psi}_0(x) = \min\{1, \psi_0\}$ satisfies $\mathcal{E}_R[\tilde{\psi}_0] \leq \mathcal{E}_R[\psi_0]$. Hence the function $\varphi_0 = 1 - \psi_0$ is nonnegative. It satisfies

$$-\Delta \varphi_0 + \tfrac{1}{2} W \varphi_0 = \tfrac{1}{2} W \tag{9.81}$$

and the Dirichlet boundary condition $\varphi_0 = 0$ for $|x| = R$.

Integration of (9.75), using that $\psi_0(r) = 1$ for $r = R$, gives

$$E_R = \tfrac{1}{2} \int W \psi_0 = \tfrac{1}{2} \int W(1 - \varphi_0). \tag{9.82}$$

Since $\varphi_0 \geq 0$ we thus need to show that $\|\varphi_0\|_\infty = o(1)$.

By (9.81), and since φ_0 and W are both nonnegative, we have $-\Delta\varphi_0 \leq \tfrac{1}{2}W$ and hence

$$\varphi_0(x) \leq \int K_0(x, x') W(x') d^2 x' \tag{9.83}$$

where $K_0(x, x')$ is the (nonnegative) integral kernel of $(-\Delta)^{-1}$ with Dirichlet boundary conditions at $|x| = R$. The kernel $K_0(x, x')$ is integrable (the singularity is $\sim \ln|x - x'|$) and hence, if W is bounded, we have $\|\varphi_0\|_\infty \leq (\text{const.})\lambda\|w\|_\infty O(\lambda)$.

If w is not bounded we can, for every $\varepsilon > 0$, find a bounded $w^\varepsilon \leq w$ with $\int(w - w^\varepsilon) \leq \varepsilon$. Define $C_\varepsilon = \|w^\varepsilon\|_\infty$. Without restriction we can assume that C_ε is monotonously increasing with ε and continuous. The function $g(\varepsilon) = \varepsilon/C_\varepsilon$ is then monotonously decreasing and continuous with $g(\varepsilon) \to 0$ if $\varepsilon \to 0$. For every (sufficiently small) λ there is an $\varepsilon(\lambda) = o(1)$ such that $g(\varepsilon(\lambda)) = \lambda$. Then

$$\|\varphi_0\|_\infty \leq (\text{const.})(\varepsilon(\lambda) + \lambda C_{\varepsilon(\lambda)}) = (\text{const.})\varepsilon(\lambda) = o(1). \tag{9.84}$$

$\square$

Corollary 9.4 (Scattering length for scaled, soft potentials). *Assume* $W_{R,\lambda}(x) = \lambda R^{-2} w_1(x/R)$ *with* $w_1 \geq 0$ *fixed and* $\int w_1 = 1$. *Then the scattering length of* $W_{R,\lambda}$ *is*

$$a_{\text{scatt}} = R \exp\left(-\frac{4\pi + \eta(\lambda)}{\lambda}\right) \tag{9.85}$$

with $\eta(\lambda) \to 0$ *for* $\lambda \to 0$, *uniformly in* R.

Proof. This follows from Lemma 9.3 noting that, by scaling, the scattering length of $W_{R,\lambda}$ is R times the scattering length of λw_1. The latter is independent of R. $\square$

Remark. If W is obtained by averaging a 3D integrable potential v over an interval of length h in the z variable, the formula (9.85), together with Eq. (C.8), motivates the exponential dependence of the effective 2D scattering length (9.14) of v on h/a: The integral $\lambda = \int W(x)d^2x$ is $h^{-1} \int v(\mathbf{x})d^3\mathbf{x}$, which for weak potentials is $h^{-1}8\pi a$ to lowest order, by Eq. (C.8). Inserting this into (9.85) gives (9.14) (apart from the dependence on the shape function s). This heuristics is, of course, only valid for soft potentials v. An essential step in the lower bound in the next section is the replacement of v by a soft potential to which this reasoning can indeed be applied.

9.4 Lower Bound

In the same way as for the upper bound we restrict the attention to the homogeneous case and finite boxes in the 2D variables, this time with Neumann boundary conditions. The optimal distribution of particles among the boxes is determined by using subadditivity and convexity arguments as in Chapters 2 and 6. Inhomogeneity in the 2D x variables can be treated by factorizing out a product of GP minimizers as mentioned at the beginning of Section 9.2.

In the treatment of the lower bound there is a natural division line between the case where the mean particle distance $\rho^{-1/2}$ is comparable to or larger than h and the case that it is much smaller than h. The first case includes Region II and a part (but not all) of Region I. When $\rho^{-1/2}$ is much smaller than h the boxes have finite extension in the z direction as well. The method is then a fairly simple modification of the 3D estimates in Chapter 6 (see also Section 4.4 in [LSeY6]) and will not be discussed further here.

The derivation of a lower bound for the case that $\rho h^2 \leq C < \infty$ proceeds by the following steps:

- Use Dyson's Lemma 2.5 to replace v by an integrable 3D potential U, retaining part of the kinetic energy.

- Average the potential U at fixed $x \in \mathbb{R}^2$ over the z-variable to obtain a 2D potential W. Estimate the error in this averaging procedure by using Temple's inequality (2.51) at each fixed x.

- The result is a 2D many body problem with an integrable interaction potential W which, by Corollary 9.4, has the right 2D scattering length to lowest order in a/h, but reduced kinetic energy inside the range of the potential. Moreover, the many body Hamiltonian has only nearest neighbor interactions. This problem is treated in the same way as in Chapter 3, introducing a 2D Dyson potential and using perturbation theory, again estimating the errors by Temple's inequality.

- Choose the parameters (size ℓ of box, fraction ε of the kinetic energy, range R of potential U, as well as the corresponding parameters for the 2D Dyson potential) optimally to minimize the errors.

The first two steps are analogous to the corresponding steps in the proof of the lower bound in Lemma 8.4, cf. Eqs. (8.48)–(8.53). It is, however, convenient to define the Dyson potential U in a slightly different manner than in Eq. (2.43).

Namely, for $R \geq 2R_0$ with R_0 the range of v we define

$$U_R(r) = \begin{cases} \frac{24}{7} R^{-3} & \text{for } \frac{1}{2}R < r < R \\ 0 & \text{otherwise.} \end{cases} \tag{9.86}$$

The reason is that this potential has a simple scaling with R which is convenient when applying Corollary 9.4. Proceeding as in Eqs. (8.48)–(8.53) we write a general wave function as

$$\Psi(\mathbf{x}_1, \ldots, \mathbf{x}_n) = f(\mathbf{x}_1, \ldots, \mathbf{x}_n) \prod_{k=1}^{n} s_h(z_k) , \tag{9.87}$$

and define $F(x_1, \ldots, x_n) \geq 0$ by

$$|F(x_1, \ldots, x_n)|^2 = \int |f(\mathbf{x}_1, \ldots, \mathbf{x}_n)|^2 \prod_{k=1}^{n} s_h(z_k)^2 dz_k . \tag{9.88}$$

Note that F is normalized if Ψ is normalized. The analogue of Eq. (8.53) is

$$\langle \Psi | H | \Psi \rangle - \frac{ne^{\perp}}{h^2} \geq$$

$$\sum_{i=1}^{n} \int \left[\varepsilon |\nabla_i^{\|} F|^2 + (1-\varepsilon) |\nabla_i^{\|} F|^2 \chi_{\min_k |x_i - x_k| \geq R}(x_i) \right] \prod_{k=1}^{n} dx_k$$

$$+ \sum_{i=1}^{n} \int \left[\varepsilon |\partial_i f|^2 + a' U_R(|\mathbf{x}_i - \mathbf{x}_{k(i)}|) \chi_{\mathcal{B}_\delta}(z_{k(i)}/h) |f|^2 \right] \prod_{k=1}^{n} s_h(z_k)^2 d\mathbf{x}_k, \tag{9.89}$$

where and $\nabla_i^{\|}$ denotes the gradient with respect to x_i and $\partial_j = d/dz_j$. Moreover, $\chi_{\mathcal{B}_\delta}$ is the characteristic function of the subset $\mathcal{B}_\delta \subset \mathbb{R}$ where $s(z)^2 \geq \delta$ for $\delta > 0$,

$$a' = a(1-\varepsilon)(1 - 2R\|\partial s^2\|_{\infty}/h\delta), \tag{9.90}$$

and $k(i)$ denotes the index of the nearest neighbor to $\mathbf{x}_i$. We now consider, for fixed $x_1, \ldots, x_n$, the term

$$\sum_{i=1}^{n} \int \left[\varepsilon |\partial_i f|^2 + a' U_R(|\mathbf{x}_i - \mathbf{x}_{k(i)}|) \chi_{\mathcal{B}_\delta}(z_{k(i)}/h) |f|^2 \right] \prod_{k=1}^{n} s_h(z_k)^2 dz_k . \tag{9.91}$$

This is estimated from below in the same way as in Eqs. (8.56), using Temple's inequality. The result is, by a calculation analogous to Eqs. (8.56)–(8.60),

$$\langle \Psi | H | \Psi \rangle - \frac{ne^{\perp}}{h^2} \geq \int \sum_{i=1}^{n} \left[\varepsilon |\nabla_i^{\|} F|^2 + (1-\varepsilon) |\nabla_i^{\|} F|^2 \chi_{|x_i - x_{k(i)}| \geq R}(x_i) \right.$$

$$\left. + \frac{1}{2} W(x_i - x_{k(i)}) |F|^2 \right] \prod_{k=1}^{n} dx_k , \tag{9.92}$$

where $x_{k(i)}$ denotes here the nearest neighbor to x_i among the points $x_k \in \mathbb{R}^2$, $k \neq i$ and W is obtained by averaging $a'U_R$ over z:

$$W(x - x') = 2a'' \int_{\mathbb{R} \times \mathbb{R}} s_h(z)^2 s_h(z')^2 U_R(|\mathbf{x} - \mathbf{x}'|) \chi_{\mathcal{B}_\delta}(z'/h) dz dz' . \tag{9.93}$$

Here, $a'' = a'(1 - \eta)$ with an error term η containing the error estimates form the Temple inequality. These errors, and the replacement of $n - 1$ by n, require the following terms to be small:

$$\varepsilon, \quad \frac{nR}{h}, \quad \frac{nh^2 a}{\varepsilon R^3}, \quad \frac{1}{n}. \tag{9.94}$$

The potential W can be written as

$$W(x) = \lambda R^{-2} w_1(x/R), \tag{9.95}$$

where w_1 is independent of R, with

$$\int w_1(x) d^2 x = 1 \tag{9.96}$$

and

$$\lambda = \frac{8\pi \, a \int s^4}{h} (1 - \eta'). \tag{9.97}$$

Here, η' is another error term involving δ and $R/h\delta$ (cf. (9.90)) besides (9.94). The 2D scattering length of (9.95) can be computed by Corollary 9.4 and has the right form (9.14) to leading order in λ. (Recall from the remark preceding Eq. (9.16) that R in (9.85) can be replaced by h as long as $ca < R < Ch$.)

The Hamiltonian on the right side of Eq. (9.92) can now be treated with the methods of Chapter 3. The only differences to the Hamiltonian discussed in that chapter are that the Hamiltonian in (9.92) has reduced kinetic energy inside the range of the potential W and the interaction is only with nearest neighbors. The latter is of no consequence since the method of Voronoi cells employed in Chapter 3 uses only this part of the interaction anyhow. The reduced kinetic energy inside the range of W implies that λ in the error term $\eta(\lambda)$ in Corr. 9.4 should be replaced by λ/ε, but otherwise the method is the same: a slight modification of Dyson's Lemma 3.1 allows to substitute for W a potential of larger range to which perturbation theory can be applied. Denoting the range of this 2D potential by $\tilde{R}$ and the fraction of the kinetic energy borrowed in the 2D Dyson Lemma by $\tilde{\varepsilon}$ the error terms to be controlled in this step are

$$\tilde{\varepsilon}, \quad \frac{R}{\tilde{R}}, \quad \frac{\tilde{R}}{\ell}, \quad \frac{n\ell^2}{\tilde{\varepsilon}\tilde{R}^2 \ln(\tilde{R}^2/a_{2D}^2)}. \tag{9.98}$$

Using superadditivity of the energy (which follows from $W \geq 0$) and convexity in the same way as in Eqs. (2.55)–(2.57) one sees that if $\rho = N/L^2$ is the density in the thermodynamically large box of side length L the optimal choice of n in the box of fixed side length ℓ is $n \sim \rho \ell^2$. We thus have to show that it is possible to choose the parameters ε, R, δ, ℓ, $\tilde{\varepsilon}$ and $\tilde{R}$ in such a way that all the error terms (9.94) and (9.98), as well as δ and $R/h\delta$ are small. We note that the conditions $a/h \ll 1$ and $\rho|\ln(\rho a_{2D}^2)|^{-1} \ll 1/h^2$ imply $\rho a_{2D}^2 \to 0$ and hence $|\ln(\rho a_{2D}^2)|^{-1} \to 0$.

We make the ansatz

$$\varepsilon = \left(\frac{a}{h}\right)^{\alpha}, \quad \delta = \left(\frac{a}{h}\right)^{\alpha'}, \quad R = h\left(\frac{a}{h}\right)^{\beta}, \quad \ell = \rho^{-1/2}\left(\frac{a}{h}\right)^{\gamma}. \tag{9.99}$$

Then $n = \rho \ell^2 = (a/h)^{-2\gamma}$ and $R/h\delta = (a/h)^{\beta - \alpha'}$. The error terms (9.94) are also powers of a/h and we have to ensure that all exponents are positive, in particular

$$\beta - \alpha' > 0, \quad \beta - 2\gamma > 0, \quad 1 - \alpha - 3\beta - 2\gamma > 0. \tag{9.100}$$

This is fulfilled, e.g., for

$$\alpha = \alpha' = \frac{1}{9}, \quad \beta = \frac{2}{9}, \quad \gamma = \frac{1}{18}, \tag{9.101}$$

with all the exponents (9.100) equal to $1/9$.

Next we write

$$\tilde{\varepsilon} = |\ln(\rho a_{2D}^2)|^{-\delta}, \quad \tilde{R} = \rho^{-1/2}|\ln(\rho a_{2D}^2)|^{-\varsigma}. \tag{9.102}$$

The error terms (9.98) are then

$$\tilde{\varepsilon} = |\ln(\rho a_{2D}^2)|^{-\delta}, \quad \frac{R}{\tilde{R}} = \left(\frac{a}{h}\right)^{\beta}(\rho h^2)^{1/2}|\ln(\rho a_{2D}^2)|^{\varsigma}, \quad \frac{\tilde{R}}{\ell} = \left(\frac{a}{h}\right)^{\gamma}|\ln(\rho a_{2D}^2)|^{-\varsigma}, \tag{9.103}$$

and

$$\frac{n\ell^2}{\tilde{\varepsilon}\tilde{R}^2\ln(\tilde{R}^2/a_{2D}^2)} = \left(\frac{a}{h}\right)^{-4\gamma}|\ln(\rho a_{2D}^2)|^{-(1-\delta-2\varsigma)}(1 + O(\ln|\ln(\rho a_{2D}^2)|)). \tag{9.104}$$

Since $(a/h)^{-4\gamma}|\ln(\rho a_{2D}^2)|^{-4\gamma} = O(1)$, the error term (9.104) can also be written as

$$\frac{n\ell^2}{\tilde{\varepsilon}\tilde{R}^2\ln(\tilde{R}^2/a_{2D}^2)} = O(1)|\ln(\rho a_{2D}^2)|^{-(1-\delta-2\varsigma-4\gamma)}(1 + O(\ln|\ln(\rho a_{2D}^2)|)). \tag{9.105}$$

The condition $\rho h^2 < C$ is used to bound $R/\tilde{R}$ in (9.103). Namely,

$$(\rho h^2)^{1/2}|\ln(\rho a_{2D}^2)|^\varsigma \leq (\text{const.})(h/a)^\varsigma, \tag{9.106}$$

so

$$\frac{R}{\tilde{R}} = O(1) \left(\frac{a}{h}\right)^{\beta - \varsigma}. \tag{9.107}$$

We choose now

$$\varsigma = \frac{1}{9}, \quad \delta = \frac{2}{9}. \tag{9.108}$$

Then

$$\beta - \varsigma = \frac{1}{9}, \quad \text{and} \quad \delta = 1 - \delta - \varsigma - 4\gamma = \frac{2}{9}. \tag{9.109}$$

This completes our discussion of the lower bound for the case $\rho h^2 \leq C < \infty$ and a homogeneous system in the 2D x variables. As already mentioned, the case $\rho h^2 \gg 1$ can be treated with the 3D methods of Chapters 2 and 6 and inhomogeneity in the x variables by separating out a product of GP minimizers.

Chapter 10

The Charged Bose Gas, the One- and Two-Component Cases

The setting now changes abruptly. Instead of particles interacting with a short-range potential $v(|\mathbf{x}_i - \mathbf{x}_j|)$ they interact via the Coulomb potential

$$v(|\mathbf{x}_i - \mathbf{x}_j|) = |\mathbf{x}_i - \mathbf{x}_j|^{-1}$$

(in 3 dimensions). The unit of electric charge is 1 in our units.

We will here consider both the one- and two-component gases. In the one-component gas (also referred to as the one-component plasma or bosonic jellium) we consider positively charged particles confined to a box with a uniformly charged background. In the two-component gas we have particles of both positive and negative charges moving in all of space.

10.1 The One-Component Gas

In the one-component gas there are N positively charged particles in a large box Λ of volume L^3 as before, with $\rho = N/L^3$.

To offset the huge Coulomb repulsion (which would drive the particles to the walls of the box) we add a uniform negative background of precisely the same charge, namely density ρ. Our Hamiltonian is thus

$$H_N^{(1)} = \sum_{i=1}^{N} -\mu\Delta_i - V(\mathbf{x}_i) + \sum_{1\leq i<j\leq N} v(|\mathbf{x}_i - \mathbf{x}_j|) + C \tag{10.1}$$

with

$$V(\mathbf{x}) = \rho \int_\Lambda |\mathbf{x} - \mathbf{y}|^{-1}d\mathbf{y} \qquad \text{and} \qquad C = \frac{1}{2}\rho \int_\Lambda V(\mathbf{x})d\mathbf{x} \ .$$

We shall use Dirichlet boundary conditions. As before the Hamiltonian acts on symmetric wave functions in $L^2(\Lambda^N, d\mathbf{x}_1 \cdots d\mathbf{x}_N)$.

Each particle interacts only with others and not with itself. Thus, despite the fact that the Coulomb potential is positive definite, the ground state energy E_0 can be (and is) negative (just take $\Psi =$const.). This time, *large ρ* is the 'weakly interacting' regime.

We know from the work in [LN] that the thermodynamic limit $e_0(\rho)$ defined as in (2.2) exists. It also follows from this work that we would, in fact, get the same thermodynamic energy if we did not restrict the number of particles N, but considered the grand-canonical case where we minimize the energy over all possible particle numbers, but keeping the background charge ρ fixed.

Another way in which this problem is different from the previous one is that *perturbation theory is correct to leading order*. If one computes $(\Psi, H\Psi)$ with $\Psi =$const, one gets the right first order answer, namely 0. It is the next order in $1/\rho$ that is interesting, and this is *entirely* due to correlations. In 1961 Foldy [F] calculated this correlation energy according to the prescription of Bogoliubov's 1947 theory. That theory was not exact for the dilute Bose gas, as we have seen, even to first order. We are now looking at *second* order, which should be even worse. Nevertheless, there was good physical intuition that this calculation should be asymptotically *exact*. Indeed it is, as proved in [LSo] and [So].

The Bogoliubov theory states that the main contribution to the energy comes from pairing of particles into momenta $\mathbf{k}, -\mathbf{k}$ and is the bosonic analogue of the BCS theory of superconductivity which came a decade later. I.e., Ψ_0 is a sum of products of terms of the form $\exp\{i\mathbf{k} \cdot (\mathbf{x}_i - \mathbf{x}_j)\}$.

The following theorem is the main result for the one-component gas.

Theorem 10.1 (Foldy's law for the one-component gas).

$$\lim_{\rho\to\infty} \rho^{-1/4}e_0(\rho) = -\frac{2}{5}\frac{\Gamma(3/4)}{\Gamma(5/4)} \left(\frac{2}{\mu\pi}\right)^{1/4}. \tag{10.2}$$

This is the *first example* (in more than 1 dimension) in which Bogoliubov's pairing theory has been rigorously validated. It has to be emphasized, however, that Foldy and Bogoliubov rely on the existence of Bose-Einstein condensation. We neither make such a hypothesis nor does our result for the energy imply the existence of such condensation. As we said earlier, it is sufficient to prove condensation in small boxes of fixed size.

Incidentally, the one-dimensional example for which Bogoliubov's theory is asymptotically exact to the first two orders (high density) is the repulsive delta-function Bose gas [LL], discussed in Appendix B, for which there is no Bose-Einstein condensation.

To appreciate the $-\rho^{1/4}$ nature of (10.2), it is useful to compare it with what one would get if the bosons had infinite mass, i.e., the first term in (10.1) is dropped. Then the energy would be proportional to $-\rho^{1/3}$ as shown in [LN]. Thus, the effect of quantum mechanics is to lower $\frac{1}{3}$ to $\frac{1}{4}$.

A problem somewhat related to bosonic jellium is *fermionic* jellium. Graf and Solovej [GS] have proved that the first two terms are what one would expect, namely

$$e_0(\rho) = C_{\mathrm{TF}}\rho^{5/3} - C_{\mathrm{D}}\rho^{4/3} + o(\rho^{4/3}), \qquad (10.3)$$

where C_{TF} is the usual Thomas-Fermi constant and C_{D} is the usual Dirac exchange constant.

It is supposedly true, for both bosonic and fermionic particles, that there is a critical mass above which the ground state should show crystalline ordering (Wigner crystal), but this has never been proved and it remains an intriguing open problem, even for the infinite mass case. A simple scaling shows that large mass is the same as small ρ, and is thus outside the region where a Bogoliubov approximation can be expected to hold.

As for the dilute Bose gas, there are several relevant length scales in the problem of the charged Bose gas. For the dilute gas there were three scales. This time there are just two. Because of the long range nature of the Coulomb problem there is no scale corresponding to the scattering length a. One relevant length scale is again the interparticle distance $\rho^{-1/3}$. The other is the correlation length scale $\ell_{\mathrm{cor}} \sim \rho^{-1/4}$ (ignoring the dependence on μ). The order of the correlation length scale can be understood heuristically as follows. Localizing on a scale ℓ_{cor} requires kinetic energy of the order of ℓ_{cor}^{-2}. The Coulomb potential from the particles and background on the scale ℓ_{cor} is $(\rho\ell_{\mathrm{cor}}^3)/\ell_{\mathrm{cor}}$. Thus the kinetic energy and the Coulomb energy balance when $\ell_{\mathrm{cor}} \sim \rho^{-1/4}$. This heuristics is however much too simplified and hides the true complexity of the situation.

Note that in the high density limit ℓ_{cor} is long compared to the interparticle distance. This is analogous to the dilute gas where the scale ℓ_c is also long compared to the interparticle distance [see (2.12)]. There is however no real analogy between the scale ℓ_{cor} for the charged gas and the scale ℓ_c for the dilute gas. In particular, whereas $e_0(\rho)$ for the dilute gas is, up to a constant, of the same order as the kinetic energy $\sim \mu\ell_c^{-2}$ we have for the charged gas that $e_0(\rho) \not\sim \ell_{\mathrm{cor}}^{-2} = \rho^{1/2}$. The reason for this difference is that on average only a small fraction of the particles in the charged gas actually correlate.

10.2   The Two-Component Gas

Now we consider N particles with charges ± 1. The Hamiltonian is thus

$$H_N^{(2)} = \sum_{i=1}^{N} -\mu\Delta_i + \sum_{1 \leq i < j \leq N} \frac{e_i e_j}{|\mathbf{x}_i - \mathbf{x}_j|}.$$

This time we are interested in $E_0^{(2)}(N)$ the ground state energy of $H_N^{(2)}$ minimized over all possible combination of charges $e_i = \pm 1$, i.e., we do not necessarily assume that the minimum occurs for the neutral case. Restricting to the neutral case would however not change the result we give below.

An equivalent formulation is to say that $E_0^{(2)}(N)$ is the ground state energy of the Hamiltonian acting on all wave functions of space and charge, i.e., functions in $L^2\left((\mathbb{R}^3 \times \{-1, 1\})^N\right)$. As mentioned in the introduction, and explained in the beginning of the proof of Thm. 2.2, for the calculation of the ground state energy we may as usual restrict to symmetric functions in this Hilbert space.

For the two-component gas there is no thermodynamic limit. In fact, Dyson [D2] proved that $E_0^{(2)}(N)$ was at least as negative as $-(\mathrm{const})N^{7/5}$ as $N \to \infty$. Thus, thermodynamic stability (i.e., a linear lower bound) fails for this gas. Years later, a lower bound of this $-N^{7/5}$ form was finally established in [CLY], thereby proving that this law is correct.

The connection of this $-N^{7/5}$ law with the jellium $-\rho^{1/4}$ law (for which a corresponding lower bound was also given in [CLY]) was pointed out by Dyson [D2] in the following way. Assuming the correctness of the $-\rho^{1/4}$ law, one can treat the 2-component gas by treating each component as a background for the other. What should the density be? If the gas has a radius L and if it has N bosons then $\rho = NL^{-3}$. However, the extra kinetic energy needed to compress the gas to this radius is NL^{-2}. The total energy is then $NL^{-2} - N\rho^{1/4}$, and minimizing this with respect to L gives $L \sim N^{-1/5}$ and leads to the $-N^{7/5}$ law. The correlation length scale is now $\ell_{\text{cor}} \sim \rho^{-1/4} \sim N^{-2/5}$.

In [D2] Dyson conjectured an exact asymptotic expression for $E_0^{(2)}(N)$ for large N. That this asymptotics, as formulated in the next theorem, is indeed correct is proved in [LSo2] and [So].

Theorem 10.2 (Dyson's law for the two-component gas).

$$\lim_{N \to \infty} \frac{E_0^{(2)}(N)}{N^{7/5}} = \inf\left\{\mu \int_{\mathbb{R}^3} |\nabla \Phi|^2 - I_0 \int_{\mathbb{R}^3} \Phi^{5/2} \,\middle|\, 0 \leq \Phi, \int_{\mathbb{R}^3} \Phi^2 = 1\right\}, \quad (10.4)$$

where I_0 is the constant from Foldy's law:

$$I_0 = \frac{2}{5}\frac{\Gamma(3/4)}{\Gamma(5/4)}\left(\frac{2}{\mu\pi}\right)^{1/4}.$$

This asymptotics can be understood as a mean field theory for the gas density, very much like the Gross-Pitaevskii functional for dilute trapped gases, where the local energy described by Foldy's law should be balanced by the kinetic energy of the gas density. Thus if we let the gas density be given by ϕ^2 then the "mean field" energy should be

$$\mu \int_{\mathbb{R}^3} |\nabla \phi|^2 - I_0 \int_{\mathbb{R}^3} \phi^{5/2}. \quad (10.5)$$

Here $\int \phi^2 = N$. If we now define $\Phi(\mathbf{x}) = N^{-8/5}\phi(N^{-1/5}\mathbf{x})$ we see that $\int \Phi^2 = 1$ and that the above energy is

$$N^{7/5}\left(\mu \int_{\mathbb{R}^3} |\nabla \Phi|^2 - I_0 \int_{\mathbb{R}^3} \Phi^{5/2}\right).$$

It may be somewhat surprising that it is exactly the same constant I_0 that appears in both the one- and two-component cases. The reason that there are no extra factors to account for the difference between one and two components is, as we shall see below, a simple consequence of Bogoliubov's method. The origin of this equivalence, while clear mathematically, does not appear to have a simple physical interpretation.

10.3   The Bogoliubov Approximation

In this section we shall briefly explain the Bogoliubov approximation and how it is applied in the case of the charged Bose gas. The Bogoliubov method relies on the exact diagonalization of a Hamiltonian, which is quadratic in creation and annihilation operators. For the charged Bose gas one only needs a very simple case of the general diagonalization procedure. On the other hand, the operators that appear are not exact creation and annihilation operators. A slightly more general formulation is needed.

Theorem 10.3 (Simple case of Bogoliubov's method). *Assume that $b_{\pm,\pm}$ are four (possibly unbounded) commuting operators satisfying the operator inequality*

$$\left[b_{\tau,e}, b^*_{\tau,e}\right] \leq 1 \quad \text{for all } e, \tau = \pm.$$

Then for all real numbers $A, B_+, B_- \geq 0$ we have

$$A \sum_{\tau,e=\pm 1} b^*_{\tau,e} b_{\tau,e}$$

$$+ \sum_{e,e'=\pm 1} \sqrt{B_e B_{e'}} \, ee' (b^*_{+,e} b_{+,e'} + b^*_{-,e} b_{-,e'} + b^*_{+,e} b^*_{-,e'} + b_{+,e} b_{-,e'})$$

$$\geq \quad -(A + B_+ + B_-) + \sqrt{(A + B_+ + B_-)^2 - (B_+ + B_-)^2}.$$

If $b_{\pm,\pm}$ are four annihilation operators then the lower bound is sharp.

Proof. Let us introduce

$$d^*_\pm = (B_+ + B_-)^{-1/2}(B_+^{1/2} b^*_{\pm,+} - B_-^{1/2} b^*_{\pm,-}),$$

and

$$c^*_\pm = (B_+ + B_-)^{-1/2}(B_-^{1/2} b^*_{\pm,+} + B_+^{1/2} b^*_{\pm,-}).$$

Then these operators satisfy

$$[d_+, d^*_+] \leq 1, \quad [d_-, d^*_-] \leq 1.$$

The operator that we want to estimate from below may be rewritten as

$$A(d^*_+ d_+ + d^*_- d_- + c^*_+ c_+ + c^*_- c_-)$$
$$+(B_+ + B_-)\left(d^*_+ d_+ + d^*_- d_- + d^*_+ d^*_- + d_+ d_-\right).$$

We may now complete the squares to write this as

$$A(c^*_+ c_+ + c^*_- c_-) + D(d^*_+ + \lambda d_-)(d^*_+ + \lambda d_-)^*$$
$$+D(d^*_- + \lambda d_+)(d^*_- + \lambda d_+)^* - D\lambda^2([d_+, d^*_+] + [d_-, d^*_-])$$

if

$$D(1 + \lambda^2) = A + B_+ + B_-, \quad 2D\lambda = B_+ + B_-.$$

We choose the solution $\lambda = 1 + \frac{A}{B_+ + B_-} - \sqrt{\left(1 + \frac{A}{(B_+ + B_-)}\right)^2 - 1}$. Hence

$$D\lambda^2 = \tfrac{1}{2}\left(A + B_+ + B_- - \sqrt{(A + B_+ + B_-)^2 - (B_+ + B_-)^2}\right). \qquad \square$$

In the theorem above one could of course also have included linear terms in $b_{\tau,e}$ in the Hamiltonian. In the technical proofs in [LSo, LSo2] the Bogoliubov diagonalization with linear terms is indeed being used to control certain error terms. Here we shall not discuss the technical details of the proofs. We have therefore stated the theorem in the simplest form in which we shall need it to derive the leading contribution.

In our applications to the charged Bose gas the operators $b_{\pm,e}$ will correspond to the annihilation of particles with charge $e = \pm$ and momenta $\pm\mathbf{k}$ for some $\mathbf{k} \in \mathbb{R}^3$. Thus, only equal and opposite momenta couple. In a translation invariant case this would be a simple consequence of momentum conservation. The one-component gas is not translation invariant, in our formulation. The two-component gas is translation invariant, but it is natural to break translation invariance by going into the center of mass frame. In both cases it is only in some approximate sense that equal and opposite momenta couple.

In the case of the one-component gas we only need particles of one sign. In this case we use the above theorem with $b_{\pm,-} = 0$ and $B_- = 0$.

We note that the lower bounds in Theorem 10.3 for the one- and two-component gases are the same except for the replacement of B_+ in the one-component case by $B_+ + B_-$ in the two-component case. In the application to the two-component gas B_+ and B_- will be proportional to the particle densities for respectively the positive or negatively charged particles. For the one-component gas B_+ is proportional to the background density.

The Bogoliubov diagonalization method cannot be immediately applied to the operators $H_N^{(1)}$ or $H_N^{(2)}$ since these operators are *not* quadratic in creation and annihilation operators. In fact, they are quartic. They have the general form

$$\sum_{\alpha,\beta} t_{\alpha\beta} a_\alpha^* a_\beta + \tfrac{1}{2} \sum_{\alpha,\beta,\mu,\nu} w_{\alpha\beta\mu\nu} a_\alpha^* a_\beta^* a_\nu a_\mu, \qquad (10.6)$$

with

$$t_{\alpha\beta} = \langle\alpha|T|\beta\rangle, \qquad w_{\alpha\beta\mu\nu} = \langle\alpha\beta|W|\mu\nu\rangle,$$

where T is the one-body part of the Hamiltonian and W is the two-body-part of the Hamiltonian.

The main step in Bogoliubov's approximation is now to assume Bose-Einstein condensation, i.e., that almost all particles are in the same one-particle state. In case of the two-component gas this means that almost half the particles are positively charged and in the same one-particle state as almost all the other half of negatively charged particles. We denote this condensate state by the index $\alpha = 0$ in the sums above. Based on the assumption of condensation Bogoliubov now argues that one may ignore all terms in the quartic Hamiltonian above which contain 3 or 4 non-zero indices and at the same time replace all creation and annihilation operators of the condensate by their expectation values. The result is a quadratic Hamiltonian (including linear terms) in the creation and annihilation with non-zero index. This Hamiltonian is of course not particle number preserving, reflecting the simple fact that particles may be created out of the condensate or annihilated into the condensate.

In Section 10.5 below it is explained how to construct trial wave functions for the one- and two-component charged gases whose expectations agree essentially with the prescription in the Bogoliubov approximation. The details appear in [So]. This will imply upper bounds on the energies corresponding to the asymptotic forms given in Theorems 10.1 and 10.2.

In [LSo, LSo2] it is proved how to make the steps in the Bogoliubov approximation rigorous as lower bounds. The main difficulty is to control the degree of condensation. As already explained it is not necessary to prove condensation in the strong sense described above. We shall only prove condensation in small boxes. Put differently, we shall not conclude that most particles are in the same one-particle state, but rather prove that most particles occupy one-particle states that look the same on short scales, i.e., that vary slowly. Here the short scale is the correlation length scale ℓ_{cor}.

10.4   The Rigorous Lower Bounds

As already mentioned we must localize into small boxes of some fixed size ℓ. This time we must require $\ell_{\text{cor}} \ll \ell$. For the one-component gas this choice is made only in order to control the degree of condensation. For the two-component gas it is required both to control the order of condensation, and also to make a local constant density approximation. The reason we can control the degree of condensation in a small box is that the localized kinetic energy has a gap above the lowest energy state. In fact, the gap is of order ℓ^{-2}. Thus if we require that ℓ is such that $N\ell^{-2}$ is much greater than the energy we may conclude that most particles are in the lowest eigenvalue state for the localized kinetic energy. We shall always choose the localized kinetic energy in such a way that the lowest eigenstate, and hence the condensate, is simply a constant function.

10.4.1 Localizing the interaction

In contrast to the dilute gas the long range Coulomb potential prevents us from simply ignoring the interaction between the small boxes. To overcome this problem we use a sliding technique first introduced in [CLY].

Theorem 10.4 (Controlling interactions by sliding). *Let χ be a smooth approximation to the characteristic function of the unit cube centered at the origin. For $\ell > 0$ and $\mathbf{z} \in \mathbb{R}^3$ let $\chi_{\mathbf{z}}(\mathbf{x}) = \chi((\mathbf{x} - \mathbf{z})/\ell)$. There exists an $\omega > 0$ depending on χ (in such a way that it tends to infinity as χ approximates the characteristic function) such that*

$$\sum_{1 \leq i < j \leq N} \frac{e_i e_j}{|\mathbf{x}_i - \mathbf{x}_j|} \geq \left(\int \chi^2 \right)^{-1} \int_{\mathbb{R}^3} \sum_{1 \leq i < j \leq N} e_i e_j w_{\ell \mathbf{z}}(\mathbf{x}_i, \mathbf{x}_j) d\mathbf{z} - \frac{N\omega}{2\ell},$$

for all $\mathbf{x}_1, \ldots \in \mathbb{R}^3$ and $e_1, \ldots = \pm 1$, where

$$w_{\mathbf{z}}(\mathbf{x}, \mathbf{y}) = \chi_{\mathbf{z}}(\mathbf{x}) Y_{\omega/\ell}(\mathbf{x} - \mathbf{y}) \chi_{\mathbf{z}}(\mathbf{y})$$

with $Y_\mu(\mathbf{x}) = |\mathbf{x}|^{-1} \exp(-\mu|\mathbf{x}|)$ being the Yukawa potential.

The significance of this result is that the two-body potential $w_{\mathbf{z}}$ is localized to the cube of size ℓ centered at $\ell \mathbf{z}$. The lower bound above is thus an integral over localized interactions sliding around with the integration parameter.

We have stated the sliding estimate in the form relevant to the two-component problem. There is an equivalent version for the one-component gas, where the sum of the particle-particle, particle-background, and background-background interactions may be bounded below by corresponding localized interactions.

Since $\ell \gg \ell_{\mathrm{cor}}$ the error in the sliding estimate is much smaller than $\omega N/\ell_{\mathrm{cor}}$, which for both the one and two-component gases is of order ω times the order of the energy. Thus, since ℓ is much bigger than ℓ_{cor}, we have room to let ω be very large, i.e., χ is close to the characteristic function.

10.4.2 Localizing the kinetic energy

Having described the technique to control the interaction between localized regions we turn next to the *localization of the kinetic energy*.

For the two-component gas this is done in two steps. As already mentioned it is natural to break the translation invariance of the two-component gas. We do this by localizing the system into a box of size $L' \gg N^{-1/5}$ (which as we saw is the expected size of the gas) as follows. By a partition of unity we can divide space into boxes of this size paying a localization error due to the kinetic energy of order $NL'^{-2} \ll N^{7/5}$. We control the interaction between these boxes using the sliding technique.

We may now argue, as follows, that the energy is smallest if all the particles are in just one box. For simplicity we give this argument for the case of two boxes. Suppose the two boxes have respective wave functions ψ and $\widetilde{\psi}$. The total energy of these two non-interacting boxes is $E + \widetilde{E}$. Now put all the particles in one box with the trial function $\Psi = \psi\widetilde{\psi}$. The fact that this function is not bosonic (i.e., it is not symmetric with respect to all the variables) is irrelevant because the true bosonic ground state energy is never greater than that of any trial state (Perron-Frobenius Theorem). The energy of Ψ is

$$E + \widetilde{E} + \iint \rho_\psi(\mathbf{x})|\mathbf{x} - \mathbf{y}|^{-1}\rho_{\widetilde{\psi}}(\mathbf{y})d\mathbf{x}d\mathbf{y},$$

where ρ_ψ and $\rho_{\widetilde{\psi}}$ are the respective *charge* densities of the states ψ and $\widetilde{\psi}$. We claim that the last Coulomb term can be made non-positive. How? If it is positive then we simply change the state $\widetilde{\psi}$ by interchanging positive and negative charges (only in $\widetilde{\psi}$ and not in ψ). The reader is reminded that we have not constrained the number of positive and negative particles but only their sum. This change in $\widetilde{\psi}$ reverses the relative charge of the states ψ and $\widetilde{\psi}$ so, by symmetry the energies E and $\widetilde{E}$ do not change, whereas the Coulomb interaction changes sign.

The localization into smaller cubes of size ℓ can however not be done by a crude partition of unity localization. Indeed, this would cost a localization error of order $N\ell^{-2}$, which as explained is required to be of much greater order than the energy.

For the one-component charged gas we may instead use a Neumann localization of the kinetic energy, as for the dilute Bose gas. If we denote by $\Delta_\ell^{(\mathbf{z})}$ the Neumann Laplacian for the cube of size ℓ centered at $\mathbf{z}$ we may, in the spirit of the sliding estimate, write the Neumann localization Laplacian in all of $\mathbb{R}^3$ as

$$-\Delta = \int -\Delta_\ell^{(\ell\mathbf{z})}d\mathbf{z}.$$

In order to write the localized kinetic energy in the same form as the localized interaction we must introduce the smooth localization χ as in Theorem 10.4. This can be achieved by ignoring the low momentum part of the kinetic energy.

More precisely, there exist $\varepsilon(\chi)$ and $s(\chi)$ such that $\varepsilon(\chi) \to 0$ and $s(\chi) \to 0$ as χ approaches the characteristic function of the unit cube and such that (see Lemma 6.1 in [LSo])

$$-\Delta_\ell^{(z)} \geq (1 - \varepsilon(\chi))\mathcal{P}_\mathbf{z}\chi_\mathbf{z}(\mathbf{x})F_{\ell s(\chi)}(-\Delta)\chi_\mathbf{z}(\mathbf{x})\mathcal{P}_\mathbf{z} \tag{10.7}$$

where $\mathcal{P}_\mathbf{z}$ denotes the projection orthogonal to constants in the cube of size ℓ centered at z and

$$F_s(u) = \frac{u^2}{u + s^{-2}}.$$

For $u \ll s^{-2}$ we have that $F_s(u) \ll u$. Hence the effect of F in the operator estimate above is to ignore the low momentum part of the Laplacian.

For the two-component gas one cannot use the Neumann localization as for the one-component gas. Using a Neumann localization ignores the kinetic energy corresponding to long range variations in the wave function and one would not get the kinetic energy term $\int \mu |\nabla\Phi|^2$ in (10.4). This is the essential difference between the one- and two-component cases. This problem is solved in [LSo2] where a new kinetic energy localization technique is developed. The idea is again to separate the high and low momentum part of the kinetic energy. The high momentum part is then localized as before, whereas the low momentum part is used to connect the localized regions by a term corresponding to a discrete Laplacian. (For details and the proof the reader is referred to [LSo2].)

Theorem 10.5 (A many body kinetic energy localization). *Let $\chi_\mathbf{z}$, $\mathcal{P}_\mathbf{z}$ and F_s be as above. There exist $\varepsilon(\chi)$ and $s(\chi)$ such that $\varepsilon(\chi) \to 0$ and $s(\chi) \to 0$ as χ approaches the characteristic function of the unit cube and such that for all normalized symmetric wave functions Ψ in $L^2((\mathbb{R}^3 \times \{-1,1\})^N)$ and all $\Omega \subset \mathbb{R}^3$ we have*

$$(1 + \varepsilon(\chi)) \left(\Psi, \sum_{i=1}^N -\Delta_i \Psi\right) \geq \int_\Omega \Big[(\Psi, \mathcal{P}_{\ell\mathbf{z}}\chi_{\ell\mathbf{z}}(\mathbf{x})F_{\ell s(\chi)}(-\Delta)\chi_{\ell\mathbf{z}(\mathbf{x})}\mathcal{P}_{\ell\mathbf{z}}\Psi)$$

$$+ \tfrac{1}{2}\ell^{-2} \sum_{\substack{\mathbf{y}\in\mathbb{Z}^3 \\ |\mathbf{y}|=1}} (S_\Psi(\ell(\mathbf{z}+\mathbf{y})) - S_\Psi(\ell\mathbf{z}))^2 \Big] d\mathbf{z}$$

$$- \text{const.}\, \ell^{-2}\text{Vol}(\Omega),$$

where

$$S_\Psi(\mathbf{z}) = \sqrt{\left(\Psi, (a_{0+}^*(\mathbf{z})a_{0+}(\mathbf{z}) + a_{0-}^*(\mathbf{z})a_{0-}(\mathbf{z}))\Psi\right) + 1} - 1$$

with $a_{0\pm}(z)$ being the annihilation of a particle of charge $\pm$ in the state given by the normalized characteristic function of the cube of size ℓ centered at $\mathbf{z}$.

The first term in the kinetic energy localization in this theorem is the same as in (10.7). The second term gives rise to a discrete Laplacian for the function $S_\Psi(\ell\mathbf{z})$, which is essentially the number of condensate particles in the cube of size ℓ centered at $\ell\mathbf{z}$. Since we will eventually conclude that most particles are in the condensate this term will after approximating the discrete Laplacian by the continuum Laplacian lead to the term $\int \mu|\nabla\phi|^2$ in (10.5). We shall not discuss this any further here.

When we apply this theorem to the two-component gas the set $\ell\Omega$ will be the box of size L' discussed above. Hence the error term $\ell^{-2}\text{Vol}(\Omega)$ will be of order $L'^3/\ell^{-5} \ll (N^{2/5}\ell)^{-5}(N^{1/5}L')^3N^{7/5}$. Thus since $\ell \gg N^{-2/5}$ we may still choose $L' \gg N^{-1/5}$, as required, and have this error term be lower order than $N^{7/5}$.

10.4.3   Controlling the degree of condensation

After now having localized the problem into smaller cubes we are ready to control the degree of condensation. We recall that the condensate state is the constant function in each cube. Let us denote by $\widehat{n}_\mathbf{z}$ the number of excited (i.e., non-condensed particles) in the box of size ℓ centered at $\mathbf{z}$. Thus for the two-component gas $\widehat{n}_\mathbf{z} + a_{0+}^*(\mathbf{z})a_{0+}(\mathbf{z}) + a_{0-}^*(\mathbf{z})a_{0-}(\mathbf{z})$ is the total number of particles in the box and a similar expression gives the particle number for the one-component gas.

As discussed above we can use the fact that the kinetic energy localized to a small box has a gap above its lowest eigenvalue to control the number of excited particles. Actually, this will show that the expectation $(\Psi, \widehat{n}_\mathbf{z}\Psi)$ is much smaller than the total number of particles in the box for any state Ψ with negative energy expectation.

One needs, however, also a good bound on $(\Psi, \widehat{n}_\mathbf{z}^2\Psi)$ to control the Coulomb interaction of the non-condensed particles. This is more difficult. In [LSo] this is not achieved directly through a bound on $(\Psi, \widehat{n}_\mathbf{z}\Psi)$ in the ground state. Rather it is proved that one may change the ground state without changing its energy very much, so that it only contains values of $\widehat{n}_\mathbf{z}$ localized close to $(\Psi, \widehat{n}_\mathbf{z}\Psi)$. The following theorem gives this very general localization technique. Its proof can be found in [LSo].

Theorem 10.6 (Localizing large matrices). *Suppose that $\mathcal{A}$ is an $N + 1 \times N + 1$ Hermitian matrix and let $\mathcal{A}^k$, with $k = 0, 1, ..., N$, denote the matrix consisting of the kth supra- and infra-diagonal of $\mathcal{A}$. Let $\psi \in \mathbf{C}^{N+1}$ be a normalized vector and set $d_k = (\psi, \mathcal{A}^k\psi)$ and $\lambda = (\psi, \mathcal{A}\psi) = \sum_{k=0}^{N} d_k$. ($\psi$ need not be an eigenvector of $\mathcal{A}$.)*

Choose some positive integer $M \leq N + 1$. Then, with M fixed, there is some $n \in [0, N + 1 - M]$ and some normalized vector $\phi \in \mathbf{C}^{N+1}$ with the property that $\phi_j = 0$ unless $n + 1 \leq j \leq n + M$ (i.e., ϕ has length M) and such that

$$(\phi, \mathcal{A}\phi) \leq \lambda + \frac{C}{M^2} \sum_{k=1}^{M-1} k^2 |d_k| + C \sum_{k=M}^{N} |d_k| \,, \tag{10.8}$$

where $C > 0$ is a universal constant. (Note that the first sum starts with $k = 1$.)

To use this theorem we start with a ground state (or approximate ground state) Ψ to the many body problem. We then consider the projections of Ψ onto the eigenspaces of $\widehat{n}_\mathbf{z}$. Since the possible eigenvalues run from 0 to N these projections span an at most $N + 1$ dimensional space.

We use the above theorem with $\mathcal{A}$ being the many body Hamiltonian restricted to this $N + 1$ dimensional subspace. Since the Hamiltonian can change the number of excited particles by at most two we see that d_k vanishes for $k \geq 3$. We shall not here discuss the estimates on d_1 and d_2 (see [LSo, LSo2]). The conclusion is that we may, without changing the energy expectation of Ψ too much, assume that the values of $\widehat{n}_\mathbf{z}$ run in an interval of length much smaller than the total number of particles. We would like to conclude that this interval is close to zero. This follows from the fact that any wave function with energy expectation close to the minimum must have an expected number of excited particles much smaller than the total number of particles.

10.4.4 The quadratic Hamiltonian

Using our control on the degree of condensation it is now possible to estimate all unwanted terms in the Hamiltonian, i.e., terms that contain 3 or more creation or annihilation operators corresponding to excited (non-condensate) states. The proof which is a rather complicated bootstrapping argument is more or less the same for the one- and two-component gases. The result, in fact, shows that we can ignore other terms too. In fact if we go back to the general form (10.6) of the Hamiltonian it turns out that we can control all quartic terms except the ones with the coefficients:

$$w_{\alpha\beta00}, \quad w_{00\alpha\beta}, \quad w_{\alpha00\beta}, \quad \text{and} \quad w_{0\alpha\beta0}.$$

To be more precise, let u_α, $\alpha = 1, \dots$ be an orthonormal basis of real functions for the subspace of functions on the cube of size ℓ centered at $\mathbf{z}$ orthogonal to constants, i.e, with vanishing average in the cube. We shall now omit the subscript

$\mathbf{z}$ and let $a_{0\pm}$ be the annihilation of a particle of charge ± 1 in the normalized constant function in the cube (i.e., in the condensate). Let $a_{\alpha\pm}$ with $\alpha \neq 0$ be the annihilation operator for a particle of charge ± 1 in the state u_α. We can then show that the main contribution to the localized energy of the two-component gas comes from the Hamiltonian

$$H_{\text{local}} = \sum_{\substack{\alpha,\beta=1 \\ e=\pm 1}}^{\infty} t_{\alpha\beta} a^*_{\alpha e} a_{\beta e}$$

$$+ \tfrac{1}{2} \sum_{\substack{\alpha,\beta=1 \\ e,e'=\pm 1}} ee' w_{\alpha\beta}(2a^*_{0e} a^*_{\alpha e'} a_{0e'} a_{\beta e} + a^*_{0e} a^*_{0e'} a_{\alpha e'} a_{\beta e} + a^*_{\alpha e} a^*_{\beta e'} a_{0e'} a_{0e}),$$

where

$$t_{\alpha\beta} = \mu(u_\alpha, \mathcal{P}_{\mathbf{z}}\chi_{\mathbf{z}}(\mathbf{x}) F_{\ell s(\chi)}(-\Delta)\chi_{\mathbf{z}}(\mathbf{x})\mathcal{P}_{\mathbf{z}} u_\beta)$$

and

$$w_{\alpha\beta} = \ell^{-3} \iint u_\alpha(\mathbf{x})\chi_{\mathbf{z}}(\mathbf{x}) Y_{\omega/\ell}(\mathbf{x} - \mathbf{y})\chi_{\mathbf{z}}(\mathbf{y}) u_\beta(\mathbf{x}) d\mathbf{x} d\mathbf{y}.$$

In H_{local} we have ignored all error terms and hence also $\varepsilon(\chi) \approx 0$ and $\int \chi^2 \approx 1$.

In the case of the one-component gas we get exactly the same local Hamiltonian, except that we have only one type of particles, i.e, we may set $a_{\alpha-} = 0$ above.

Let $\nu_\pm = \sum_{\alpha=0}^{\infty} a^*_{\alpha\pm} a_{\alpha\pm}$ be the total number of particles in the box with charge ± 1. For $\mathbf{k} \in \mathbb{R}^3$ we let $\chi_{\mathbf{k},\mathbf{z}}(\mathbf{x}) = \chi_{\mathbf{z}}(\mathbf{x})e^{i\mathbf{k}\cdot\mathbf{x}}$. We then introduce the operators

$$b_{\mathbf{k}\pm} = (\ell^3 \nu_\pm)^{-1/2} a_\pm(\mathcal{P}_{\mathbf{z}}\chi_{\mathbf{k},\mathbf{z}}) a^*_{0\pm},$$

where $a_\pm(\mathcal{P}_{\mathbf{z}}\chi_{\mathbf{k},\mathbf{z}}) = \sum_{\alpha=1}^{\infty} (\chi_{\mathbf{k},\mathbf{z}}, u_\alpha) a_{\alpha\pm}$ annihilates a particle in the state $\chi_{\mathbf{k},\mathbf{z}}$ with charge ± 1. It is then clear that the operators $b_{\mathbf{k}\pm}$ all commute and a straightforward calculation shows that

$$[b_{\mathbf{k}\pm}, b^*_{\mathbf{k}\pm}] \leq (\ell^3 \nu_\pm)^{-1} \|\mathcal{P}_{\mathbf{z}}\chi_{\mathbf{z}}\|^2 a^*_{0\pm} a_{0\pm} \leq 1.$$

If we observe that

$$\sum_{\substack{\alpha,\beta=1 \\ e=\pm 1}}^{\infty} t_{\alpha\beta} a^*_{\alpha e} a_{\beta e} = (2\pi)^{-3} \int \mu F_{\ell s(\chi)}(\mathbf{k}^2) \sum_{e=\pm} a_e(\mathcal{P}_{\mathbf{z}}\chi_{\mathbf{k},\mathbf{z}})^* a_e(\mathcal{P}_{\mathbf{z}}\chi_{\mathbf{k},\mathbf{z}}) d\mathbf{k}$$

$$\geq (2\pi)^{-3}\ell^3 \int \mu F_{\ell s(\chi)}(\mathbf{k}^2) \sum_{e=\pm} b^*_{\mathbf{k}e} b_{\mathbf{k}e},$$

we see that

$$H_{\text{local}} \geq \tfrac{1}{2}(2\pi)^{-3} \int \mu \ell^3 F_{\ell s(\chi)}(\mathbf{k}^2) \sum_{e=\pm} (b_{\mathbf{k}e}^* b_{\mathbf{k}e} + b_{-\mathbf{k}e}^* b_{-\mathbf{k}e})$$

$$+ \sum_{ee'=\pm} \widehat{Y}_{\omega/\ell}(\mathbf{k}) \sqrt{\nu_e \nu_{e'}}\, ee'(b_{\mathbf{k}e}^* b_{\mathbf{k},e'} + b_{-\mathbf{k}e}^* b_{-\mathbf{k},e'} + b_{\mathbf{k}e}^* b_{-\mathbf{k},e'}^* + b_{-\mathbf{k}e} b_{\mathbf{k},e'})d\mathbf{k}$$

$$- \sum_{\alpha\beta=1} w_{\alpha\beta}(a_{\alpha+}^* a_{\beta+} + a_{\alpha-}^* a_{\beta-}).$$

The last term comes from commuting $a_{0\pm}^* a_{0\pm}$ to $a_{0\pm} a_{0\pm}^*$. It is easy to see that this last term is a bounded operator with norm bounded by

$$\text{const.}\,(\nu_+ + \nu_-)\ell^{-3}\|\widehat{Y}_{\omega/\ell}\|_\infty \leq \text{const.}\,\omega^{-2}(\nu_+ + \nu_-)\ell^{-1}.$$

When summing over all boxes we see that the last term above gives a contribution bounded by $\text{const.}\,\omega^{-2}N\ell^{-1} = \omega^{-2}(N^{2/5}\ell)^{-1}N^{7/5}$ which is lower order than the energy.

The integrand in the lower bound on H_{local} is precisely an operator of the form treated in the Bogoliubov method Theorem 10.3. Thus up to negligible errors we see that the operator H_{local} is bounded below by

$$\tfrac{1}{2}(2\pi)^{-3} \int -(\mathcal{A}(\mathbf{k}) + \mathcal{B}(\mathbf{k})) + \sqrt{(\mathcal{A}(\mathbf{k}) + \mathcal{B}(\mathbf{k}))^2 - \mathcal{B}(\mathbf{k})^2}\, d\mathbf{k},$$

where

$$\mathcal{A}(\mathbf{k}) = \mu \ell^3 F_{\ell s(\chi)}(\mathbf{k}^2) \quad \text{and} \quad \mathcal{B}(\mathbf{k}) = \nu \widehat{Y}_{\omega/\ell}(\mathbf{k})$$

with $\nu = \nu_+ + \nu_-$ being the total number of particles in the small box. A fairly simple analysis of the above integral shows that we may to leading order replace $\mathcal{A}$ by $\mu \ell^3 \mathbf{k}^2$ and $\mathcal{B}(\mathbf{k})$ by $4\pi\nu|\mathbf{k}|^{-2}$, i.e., we may ignore the cut-offs. The final conclusion is that the local energy is given to leading order by

$$\frac{-1}{2(2\pi)^3} \int 4\pi\nu|\mathbf{k}|^{-2} + \mu \ell^3 |\mathbf{k}|^2 - \sqrt{(4\pi\nu|\mathbf{k}|^{-2} + \mu\ell^3|\mathbf{k}|^2)^2 - (4\pi\nu|\mathbf{k}|^{-2})^2}\, d\mathbf{k}$$

$$= -2^{1/2}\pi^{-3/4}\nu \left(\frac{\nu}{\mu\ell^3}\right)^{1/4} \int_0^\infty 1 + x^4 - x^2(2 + x^4)^{1/2}\, dx.$$

If we finally use that

$$\int_0^\infty 1 + x^4 - x^2(2 + x^4)^{1/2}\, dx = \frac{2^{3/4}\sqrt{\pi}\,\Gamma(3/4)}{5\Gamma(5/4)}$$

we see that the local energy to leading order is $-I_0\nu(\nu/\ell^3)^{1/4}$. For the one-component gas we should set $\nu = \rho\ell^3$ and for the two-component gas we should set $\nu = \phi^2 \ell^3$ (see (10.5)). After replacing the sum over boxes by an integral and at the same time replace the discrete Laplacian by a continuum Laplacian, as described above, we arrive at asymptotic lower bounds as in Theorems 10.1 and 10.2.

There is one issue that we have not discussed at all and which played an important role in the treatment of the dilute gas. How do we know the number of particles in each of the small cubes? For the dilute gas a superadditivity argument was used to show that there was an equipartition of particles among the smaller boxes. Such an argument cannot be used for the charged gas. For the one-component gas one simply minimizes the energy over all possible particle numbers in each little box. It turns out that charge neutrality is essentially required for the energy to be minimized. Since the background charge in each box is fixed this fixes the particle number.

For the two-component there is a-priori nothing that fixes the particle number in each box. More precisely, if we ignored the kinetic energy between the small boxes it would be energetically favorable to put all particles in one small box. It is the kinetic energy between boxes, i.e., the discrete Laplacian term in Theorem 10.5, that prevents this from happening. Thus we could in principle again minimize over all particle numbers and hope to prove the correct particle number dependence (i.e., Foldy's law) in each small box. This is essentially what is done except that boxes with very many or very few particles must be treated somewhat differently from the "good" boxes. In the "bad" boxes we do not prove Foldy's law, but only weaker estimates that are adequate for the argument.

10.5   The Rigorous Upper Bounds

10.5.1   The upper bound for the two-component gas

To prove an upper bound on the energy $E_0^{(2)}(N)$ of the form given in Dyson's formula Theorem 10.2 we shall construct a trial function from the prescription in the Bogoliubov approximation. We shall use as an input a minimizer Φ for the variational problem on the right side of (10.4). That minimizers exist can be easily seen using spherical decreasing rearrangements. It is however not important that a minimizer exists. An approximate minimizer would also do for the argument given here. Define $\phi_0(\mathbf{x}) = N^{3/10}\Phi(N^{1/5}\mathbf{x})$. Then again $\int \phi_0^2 = 1$. In terms of the unscaled function ϕ in (10.5), $\phi_0(\mathbf{x}) = N^{-1}\phi(\mathbf{x})$.

Let ϕ_α, $\alpha = 1, \ldots$ be an orthonormal family of real functions all orthogonal to ϕ_0. We choose these functions below.

We follow Dyson [D2] and choose a trial function which does not have a specified particle number, i.e., a state in the bosonic Fock space.

As our trial many-body wave function we now choose

$$\Psi = \exp\left(-\lambda_0^2 + \lambda_0 a_{0+}^* + \lambda_0 a_{0-}^*\right)$$
$$\times \prod_{\alpha \neq 0} (1 - \lambda_\alpha^2)^{1/4} \exp\left(-\sum_{e,e'=\pm 1} \sum_{\alpha \neq 0} \frac{\lambda_\alpha}{4} ee' a_{\alpha,e}^* a_{\alpha,e'}^*\right) |0\rangle, \quad (10.9)$$

where $a_{\alpha,e}^*$ is the creation of a particle of charge $e = \pm 1$ in the state ϕ_α, $|0\rangle$ is the vacuum state, and the coefficients $\lambda_0, \lambda_1, \ldots$ will be chosen below satisfying $0 < \lambda_\alpha < 1$ for $\alpha \neq 0$.

It is straightforward to check that Ψ is a normalized function.

Dyson used a very similar trial state in [D2], but in his case the exponent was a purely quadratic expression in creation operators, whereas the one used here is only quadratic in the creation operators $a_{\alpha e}^*$, with $\alpha \neq 0$ and linear in $a_{0\pm}^*$. As a consequence our state will be more sharply localized around the mean of the particle number.

In fact, the above trial state is precisely what is suggested by the Bogoliubov approximation. To see this note that one has

$$(a_{0\pm} - \lambda_0)\Psi = 0, \quad \text{and} \quad \left(a_{\alpha+}^* - a_{\alpha-}^* + \lambda_\alpha(a_{\alpha+} - a_{\alpha-})\right)\Psi = 0$$

for all $\alpha \neq 0$. Thus the creation operators for the condensed states can be replaced by their expectation values and an adequate quadratic expression in the non-condensed creation and annihilation operators is minimized.

Consider now the operator

$$\gamma = \sum_{\alpha=1}^{\infty} \frac{\lambda_\alpha^2}{1 - \lambda_\alpha^2} |\phi_\alpha\rangle\langle\phi_\alpha|. \quad (10.10)$$

A straightforward calculation of the energy expectation in the state Ψ gives that

$$\left(\Psi, \sum_{N=0}^{\infty} H_N^{(2)} \Psi\right) = 2\lambda_0^2 \mu \int (\nabla\phi_0)^2 + \text{Tr}\left(-\mu\Delta\gamma\right)$$
$$+ 2\lambda_0^2 \text{Tr}\left(\mathcal{K}\left(\gamma - \sqrt{\gamma(\gamma+1)}\right)\right),$$

where $\mathcal{K}$ is the operator with integral kernel

$$\mathcal{K}(\mathbf{x}, \mathbf{y}) = \phi_0(\mathbf{x})|\mathbf{x} - \mathbf{y}|^{-1}\phi_0(\mathbf{y}). \quad (10.11)$$

Moreover, the expected particle number in the state Ψ is $2\lambda_0^2 + \text{Tr}(\gamma)$. In order for Ψ to be well defined by the formula (10.9) we must require this expectation to be finite.

Instead of making explicit choices for the individual functions ϕ_α and the coefficients λ_α, $\alpha \neq 0$ we may equivalently choose the operator γ. In defining γ we use the method of coherent states. Let χ be a non-negative real and smooth function supported in the unit ball in $\mathbb{R}^3$, with $\int \chi^2 = 1$. Let as before $N^{-2/5} \ll \ell \ll N^{-1/5}$ and define $\chi_\ell(\mathbf{x}) = \ell^{-3/2}\chi(\mathbf{x}/\ell)$. We choose

$$\gamma = (2\pi)^{-3} \int_{\mathbb{R}^3 \times \mathbb{R}^3} f(\mathbf{u}, |\mathbf{p}|) \mathcal{P}_{\phi_0}^\perp |\theta_{\mathbf{u},\mathbf{p}}\rangle \langle \theta_{\mathbf{u},\mathbf{p}}| \mathcal{P}_{\phi_0}^\perp d\mathbf{u}d\mathbf{p}$$

where $\mathcal{P}_{\phi_0}^\perp$ is the projection orthogonal to ϕ_0,

$$\theta_{\mathbf{u},\mathbf{p}}(x) = \exp(i\mathbf{p} \cdot \mathbf{x})\chi_\ell(\mathbf{x} - \mathbf{u}),$$

and

$$f(\mathbf{u}, |\mathbf{p}|) = \frac{1}{2}\left(\frac{\mathbf{p}^4 + 16\pi\lambda_0^2\mu^{-1}\phi_0(\mathbf{u})^2}{\mathbf{p}^2\left(\mathbf{p}^4 + 32\pi\lambda_0^2\mu^{-1}\phi_0(\mathbf{u})^2\right)^{1/2}} - 1\right).$$

We note that γ is a positive trace class operator, $\gamma\phi_0 = 0$, and that all eigenfunctions of γ may be chosen real. These are precisely the requirements needed in order for γ to define the orthonormal family ϕ_α and the coefficients λ_α for $\alpha \neq 0$.

We use the following version of the Berezin-Lieb inequality [Be, L4]. Assume that $\xi(t)$ is an *operator* concave function of $\mathbb{R}_+ \cup \{0\}$ with $\xi(0) \geq 0$. Then if Y is a positive semi-definite operator we have

$$\text{Tr}\,(Y\xi(\gamma)) \geq (2\pi)^{-3} \int \xi(f(\mathbf{u}, |\mathbf{p}|))\left(\theta_{\mathbf{u},\mathbf{p}}, \mathcal{P}_{\phi_0}^\perp Y \mathcal{P}_{\phi_0}^\perp \theta_{\mathbf{u},\mathbf{p}}\right) d\mathbf{u}d\mathbf{p}. \qquad (10.12)$$

We use this for the function $\xi(t) = \sqrt{t(t+1)}$. Of course, if ξ is the identity function then (10.12) is an identity. If $Y = I$ then (10.12) holds for all concave ξ with $\xi(0) \geq 0$.

Proving an upper bound on the energy expectation (10.11) is thus reduced to the calculations of explicit integrals. After estimating these integrals one arrives at the leading contribution (for large λ_0)

$$2\lambda_0^2\mu \int (\nabla\phi_0)^2 + \iint \left(\mu\mathbf{p}^2 + 2\lambda_0^2\phi_0(\mathbf{u})^2\frac{4\pi}{\mathbf{p}^2}\right) f(\mathbf{u}, |\mathbf{p}|)$$

$$-\frac{4\pi}{\mathbf{p}^2}2\lambda_0^2\phi_0(\mathbf{u})^2\sqrt{f(\mathbf{u}, |\mathbf{p}|)(f(\mathbf{u}, |\mathbf{p}|) + 1)}\,d\mathbf{p}d\mathbf{u}$$

$$= 2\lambda_0^2\mu \int (\nabla\phi_0)^2 - I_0 \int (2\lambda_0^2)^{5/4}\phi_0^{5/2},$$

where I_0 is as in Theorem 10.2.

If we choose $\lambda_0 = \sqrt{N/2}$ we get after a simple rescaling that the energy above is $N^{7/5}$ times the right side of (10.4) (recall that Φ was chosen as the minimizer). We also note that the expected number of particles is

$$2\lambda_0^2 + \operatorname{Tr}(\gamma) = N + O(N^{3/5}),$$

as $N \to \infty$.

The only remaining problem is to show how a similar energy could be achieved with a wave function with a fixed number of particles N, i.e., how to show that we really have an upper bound on $E_0^{(2)}(N)$. We indicate this fairly simple argument here.

We construct a trial function Ψ' as above, but with an expected particle number N' chosen appropriately close to, but slightly smaller than N. More precisely, N' will be smaller than N by an appropriate lower order correction. It is easy to see then that the mean deviation of the particle number distribution in the state Ψ' is lower order than N. In fact, it is of order $\sqrt{N'} \sim \sqrt{N}$. Using that we have a good lower bound on the energy $E_0^{(2)}(n)$ for all n and that Ψ' is sharply localized around its mean particle number, we may, without changing the energy expectation significantly, replace Ψ' by a normalized wave function Ψ that only has particle numbers less than N. Since the function $n \mapsto E_0^{(2)}(n)$ is a decreasing function we see that the energy expectation in the state Ψ is, in fact, an upper bound to $E_0^{(2)}(N)$.

10.5.2 The upper bound for the one-component gas

The upper bound for the one-component gas is proved in a very similar way as for the two-component gas. We shall simply indicate the main differences here. We will again choose a trial state without a fixed particle number, i.e., a grand-canonical trial state. Since we know that the one-component gas has a thermodynamic limit and that there is equivalence of ensembles [LN], it makes no difference whether we choose a canonical or grand-canonical trial state.

For the state ϕ_0 we now choose a normalized function with compact support in Λ, that is constant on the set $\{x \in \Lambda \mid \operatorname{dist}(x, \partial\Lambda) > r\}$. We shall choose $r > 0$ to go to zero as $L \to \infty$. Let us also choose the constant n such that $n\phi_0^2 = \rho$ on the set where ϕ_0 is constant. Then $n \approx \rho L^3$.

Let again ϕ_α, $\alpha = 1, \ldots$ be an orthonormal family of real functions orthogonal to ϕ_0. As our trial state we choose, this time,

$$\Psi = \prod_{\alpha \neq 0}(1 - \lambda_\alpha^2)^{1/4} \exp\left(-\lambda_0^2/2 + \lambda_0 a_0^* - \sum_{\alpha \neq 0}\frac{\lambda_\alpha}{2}a_\alpha^* a_\alpha^*\right)|0\rangle, \quad (10.13)$$

where a_α^* is the creation of a particle in the state ϕ_α. We will choose Ψ implicitly by choosing the operator γ defined as in (10.10).

This time we obtain

$$\left(\Psi, \sum_{N=0}^{\infty} H_N^{(1)}\Psi\right) = \lambda_0^2\mu \int (\nabla\phi_0)^2$$

$$+\frac{1}{2}\iint \frac{|\gamma(\mathbf{x},\mathbf{y})|^2}{|\mathbf{x}-\mathbf{y}|}d\mathbf{x}d\mathbf{y} + \frac{1}{2}\iint \frac{|\sqrt{\gamma(\gamma+1)}(\mathbf{x},\mathbf{y})|^2}{|\mathbf{x}-\mathbf{y}|}d\mathbf{x}d\mathbf{y}$$

$$+\frac{1}{2}\iint_{\Lambda\times\Lambda} \left(\rho - \rho_\gamma(\mathbf{x}) - \lambda_0^2\phi_0(\mathbf{x})^2\right)|\mathbf{x}-\mathbf{y}|^{-1}\left(\rho - \rho_\gamma(\mathbf{y}) - \lambda_0^2\phi_0(\mathbf{y})^2\right)d\mathbf{x}d\mathbf{y}$$

$$+\text{Tr}\left(-\mu\Delta\gamma\right) + \lambda_0^2\text{Tr}\left(\mathcal{K}\left(\gamma - \sqrt{\gamma(\gamma+1)}\right)\right), \quad (10.14)$$

where $\rho_\gamma(\mathbf{x}) = \gamma(\mathbf{x},\mathbf{x})$ and $\mathcal{K}$ is again given as in (10.11). We must show that we can make choices such that the first four terms on the right side above are lower order than the energy, and can therefore be neglected.

We choose

$$\gamma = \gamma_\varepsilon = (2\pi)^{-3}\int_{|p|>\varepsilon\rho^{1/4}} f(|\mathbf{p}|)\mathcal{P}_{\phi_0}^\perp|\theta_\mathbf{p}\rangle\langle\theta_\mathbf{p}|\mathcal{P}_{\phi_0}^\perp d\mathbf{p},$$

where $\varepsilon > 0$ is a parameter which we will let tend to 0 at the end of the calculation. Here $\mathcal{P}_{\phi_0}^\perp$ as before is the projection orthogonal to ϕ_0 and this time

$$f(|\mathbf{p}|) = \frac{1}{2}\left(\frac{\mathbf{p}^4 + 8\pi\mu^{-1}\rho}{\mathbf{p}^2\left(\mathbf{p}^4 + 16\pi\mu^{-1}\rho\right)^{1/2}} - 1\right)$$

and

$$\theta_\mathbf{p}(\mathbf{x}) = \sqrt{n\rho^{-1}}\exp(i\mathbf{p}\cdot\mathbf{x})\phi_0(\mathbf{x}).$$

Note that $n\rho^{-1}\phi_0(\mathbf{x})^2$ is 1 on most of Λ. We then again have the Berezin-Lieb inequality as before. We also find that

$$\rho_\gamma(\mathbf{x}) = (2\pi)^{-3}\int_{|p|>\varepsilon\rho^{1/4}} f(|\mathbf{p}|)d\mathbf{p}n\rho^{-1}\phi_0(\mathbf{x})^2\left(1 + O(\varepsilon^{-1}\rho^{-1/4}L^{-1})\right)$$

$$= A_\varepsilon(\rho/\mu)^{3/4}n\rho^{-1}\phi_0(\mathbf{x})^2\left(1 + O(\varepsilon^{-1}\rho^{-1/4}L^{-1})\right),$$

where A_ε is an explicit function of ε. We now choose λ_0 such that $\lambda_0^2 = n(1 - A_\varepsilon \rho^{-1/4} \mu^{-3/4})$, i.e., such that

$$\lambda_0^2 \phi_0^2(\mathbf{x}) + \rho_\gamma(\mathbf{x}) = n\phi_0(\mathbf{x})^2(1 + O(\varepsilon^{-1}\rho^{-1/2}L^{-1})) \approx \rho.$$

It is easy to see that the first term in (10.14) is of order $\rho L^3(rL)^{-1}$ and the fourth term in (10.14) is of order $\rho L^3(\varepsilon^{-2} + \rho r^2)$. We may choose r, depending on L, in such a way that after dividing by ρL^3 and letting $L \to \infty$ only the error ε^{-2} remains. This allows choosing $\varepsilon \ll \rho^{-1/8}$.

To estimate the second term in (10.14) we use Hardy's inequality to deduce

$$\iint \frac{|\gamma(\mathbf{x}, \mathbf{y})|^2}{|\mathbf{x} - \mathbf{y}|} d\mathbf{x} d\mathbf{y} \leq 2(\operatorname{tr} \gamma^2)^{1/2} \operatorname{tr}(-\Delta \gamma^2)^{1/2},$$

and these terms can be easily estimated using the Berezin-Lieb inequality in the direction opposite from before, since we are interested now in an upper bound. The third term in (10.14) is controlled in exactly the same way as the second term. We are then left with the last two terms in (10.14). They are treated in exactly the same way as for the two-component gas again using the Berezin-Lieb inequality.

Chapter 11

Bose-Einstein Quantum Phase Transition in an Optical Lattice Model

11.1 Introduction

One of the most remarkable recent developments in the study of ultracold Bose gases is the observation of a reversible transition from a Bose-Einstein condensate to a state composed of localized atoms as the strength of a periodic, optical trapping potential is varied [G2, G3]. This is an example of a quantum phase transition [Sa] where quantum fluctuations and correlations rather than energy-entropy competition is the driving force and its theoretical understanding is quite challenging. The model usually considered for describing this phenomenon is the Bose-Hubbard model and the transition is interpreted as a transition between a superfluid and a *Mott insulator* that was studied in [FWGF] with an application to He^4 in porous media in mind. The possibility of applying this scheme to gases of alkali atoms in optical traps was first realized in [JBCGZ]. The article [Zw] reviews these developments and many recent papers, e.g., [Ga, Z1, NS, Ge, Z2, DODS, R, MA, AA] are devoted to this topic. These papers contain also further references to earlier work along these lines.

The investigations of the phase transition in the Bose-Hubbard model are mostly based on variational or numerical methods and the signal of the phase transition is usually taken to be that an ansatz with a sharp particle number at each lattice site leads to a lower energy than a delocalized Bogoliubov state. On the other hand, there exists no rigorous proof, so far, that the true ground state of the model has off-diagonal long range order at one end of the parameter regime that disappears at the other end. In this chapter, which is based on the paper [ALSSY], we study a slightly different model where just this phenomenon can be rigorously proved and which, at the same time, captures the salient features of the experimental situation.

Physically, we are dealing with a trapped Bose gas with short range interaction like in Chapters 1–8. The model we discuss, however, is not a continuum model but rather a lattice gas, i.e., the particles are confined to move on a d-dimensional, hypercubic lattice and the kinetic energy is given by the discrete Laplacian. Moreover, when discussing BEC, it is convenient not to fix the particle number but to work in a grand-canonical ensemble. The chemical potential is fixed in such a way that the average particle number equals half the number of lattice sites, i.e., we consider *half filling*. (This restriction is dictated by our method of proof.) The optical lattice is modeled by a periodic, one-body potential. In experiments the gas is enclosed in an additional trap potential that is slowly varying on the scale of the optical lattice but we neglect here the inhomogeneity due to such a potential and consider instead the thermodynamic limit.

In terms of bosonic creation and annihilation operators, $a_{\mathbf{x}}^{\dagger}$ and $a_{\mathbf{x}}$, our Hamiltonian is expressed as

$$H = -\tfrac{1}{2}\sum_{\langle \mathbf{xy}\rangle}(a_{\mathbf{x}}^{\dagger}a_{\mathbf{y}} + a_{\mathbf{x}}a_{\mathbf{y}}^{\dagger}) + \lambda\sum_{\mathbf{x}}(-1)^{\mathbf{x}}a_{\mathbf{x}}^{\dagger}a_{\mathbf{x}} + U\sum_{\mathbf{x}}a_{\mathbf{x}}^{\dagger}a_{\mathbf{x}}(a_{\mathbf{x}}^{\dagger}a_{\mathbf{x}} - 1). \quad (11.1)$$

The sites $\mathbf{x}$ are in a cube $\Lambda \subset \mathbb{Z}^d$ with opposite sides identified (i.e., a d-dimensional torus) and $\langle \mathbf{xy}\rangle$ stands for pairs of nearest neighbors. Units are chosen such that $\hbar^2/m = 1$.

The first term in (11.1) is the discrete Laplacian $\sum_{\langle \mathbf{xy}\rangle}(a_{\mathbf{x}}^{\dagger} - a_{\mathbf{y}}^{\dagger})(a_{\mathbf{x}} - a_{\mathbf{y}})$ minus $2d\sum_{\mathbf{x}}a_{\mathbf{x}}^{\dagger}a_{\mathbf{x}}$, i.e., we have subtracted a chemical potential that equals d.

The optical lattice gives rise to a potential $\lambda(-1)^{\mathbf{x}}$ which alternates in sign between the A and B sublattices of even and odd sites. The inter-atomic on-site repulsion is U, but we consider here only the case of a *hard core interaction*, i.e., $U = \infty$. If $\lambda = 0$ but $U < \infty$ we have the Bose-Hubbard model. Then all sites are equivalent and the lattice represents the attractive sites of the optical lattice. In our case the adjustable parameter is λ instead of U and for large λ the atoms will try to localize on the B sublattice. The Hamiltonian (11.1) conserves the particle number N and it can be shown that, for $U = \infty$, the lowest energy is obtained uniquely for $N = \tfrac{1}{2}|\Lambda|$, i.e., half the number of lattice sites. Because of the periodic potential the unit cell in this model consists of two lattice sites, so that we have on average one particle per unit cell. This corresponds, physically, to filling factor 1 in the Bose-Hubbard model.

In contrast to the previous chapters we no longer restrict our attention to the ground state of the system but consider more generally thermal equilibrium states at some nonnegative temperature T. These states are described by the Gibbs density matrices $Z^{-1}\exp(-\beta H)$ with Z the normalization factor (partition function) and $\beta = 1/T$ the inverse temperature. Units are chosen so that Boltzmann's constant equals 1. The thermal expectation value of some observable $\mathcal{O}$ will be denoted by $\langle \mathcal{O} \rangle = Z^{-1}\operatorname{tr}\mathcal{O}\exp(-\beta H)$.

Our main results about this model can be summarized as follows:

1. If T and λ are both small, there is Bose-Einstein condensation. In this parameter regime the one-body density matrix $\gamma(\mathbf{x}, \mathbf{y}) = \langle a_{\mathbf{x}}^{\dagger} a_{\mathbf{y}} \rangle$ has exactly one large eigenvalue (in the thermodynamic limit), and the corresponding condensate wave function is $\phi(\mathbf{x}) =$constant.

2. If either T or λ is big enough, then the one-body density matrix decays exponentially with the distance $|\mathbf{x} - \mathbf{y}|$, and hence there is *no BEC*. In particular, this applies to the ground state $T = 0$ for λ big enough, where the system is in a Mott insulator phase.

3. The Mott insulator phase is characterized by a gap, i.e., a jump in the chemical potential. We are able to prove this, at half-filling, in the region described in item 2 above. More precisely, there is a cusp in the dependence of the ground state energy on the number of particles; adding or removing one particle costs a non-zero amount of energy. We also show that there is no such gap whenever there is BEC.

4. The interparticle interaction is essential for items 2 and 3. Non-interacting bosons *always display BEC* for low, but positive T (depending on λ, of course).

5. For all $T \geq 0$ and all $\lambda > 0$ the diagonal part of the one-body density matrix $\langle a_{\mathbf{x}}^{\dagger} a_{\mathbf{x}} \rangle$ (the one-particle density) is *not constant*. Its value on the A sublattice is constant, but strictly less than its constant value on the B sublattice and this discrepancy survives in the thermodynamic limit. In contrast, in the regime mentioned in item 1, the off-diagonal long-range order is constant, i.e., $\langle a_{\mathbf{x}}^{\dagger} a_{\mathbf{y}} \rangle \approx \phi(\mathbf{x})\phi(\mathbf{y})^*$ for large $|\mathbf{x} - \mathbf{y}|$ with $\phi(\mathbf{x}) =$constant.

Because of the hard-core interaction between the particles, there is at most one particle at each site and our Hamiltonian (with $U = \infty$) thus acts on the

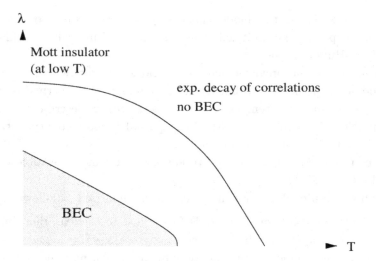

Figure 11.1: Schematic phase diagram at half-filling

Hilbert space $\mathcal{H} = \bigotimes_{x \in \Lambda} \mathbb{C}^2$. The creation and annihilation operators can be represented as 2×2 matrices with

$$a_x^\dagger \leftrightarrow \begin{pmatrix} 0 & 1 \\ 0 & 0 \end{pmatrix}, \quad a_x \leftrightarrow \begin{pmatrix} 0 & 0 \\ 1 & 0 \end{pmatrix}, \quad a_x^\dagger a_x \leftrightarrow \begin{pmatrix} 1 & 0 \\ 0 & 0 \end{pmatrix},$$

for each $x \in \Lambda$. More precisely, these matrices act on the tensor factor associated with the site x while $a_x^\dagger$ and a_x act as the identity on the other factors in the Hilbert space $\mathcal{H} = \bigotimes_{x \in \Lambda} \mathbb{C}^2$.

The Hamiltonian can alternatively be written in terms of the spin 1/2 operators

$$S^1 = \frac{1}{2} \begin{pmatrix} 0 & 1 \\ 1 & 0 \end{pmatrix}, \quad S^2 = \frac{1}{2} \begin{pmatrix} 0 & -i \\ i & 0 \end{pmatrix}, \quad S^3 = \frac{1}{2} \begin{pmatrix} 1 & 0 \\ 0 & -1 \end{pmatrix}.$$

The correspondence with the creation and annihilation operators is

$$a_x^\dagger = S_x^1 + i S_x^2 \equiv S_x^+, \quad a_x = S_x^1 - i S_x^2 \equiv S_x^-,$$

and hence $a_x^\dagger a_x = S_x^3 + \frac{1}{2}$. (This is known as the Matsubara-Matsuda correspondence [MM].) Adding a convenient constant to make the periodic potential positive, the Hamiltonian (11.1) for $U = \infty$ is thus equivalent to

$$
\begin{aligned}
H &= -\tfrac{1}{2} \sum_{\langle xy \rangle} (S_x^+ S_y^- + S_x^- S_y^+) + \lambda \sum_x \left[\tfrac{1}{2} + (-1)^x S_x^3 \right] \\
&= -\sum_{\langle xy \rangle} (S_x^1 S_y^1 + S_x^2 S_y^2) + \lambda \sum_x \left[\tfrac{1}{2} + (-1)^x S_x^3 \right].
\end{aligned}
\tag{11.2}
$$

Without loss of generality we may assume $\lambda \geq 0$. This Hamiltonian is well known as a model for interacting spins, referred to as the XY model [DLS]. The last term has the interpretation of a staggered magnetic field. We note that BEC for the lattice gas is equivalent to off-diagonal long range order for the 1- and 2-components of the spins.

The Hamiltonian (11.2) is clearly invariant under simultaneous rotations of all the spins around the 3-axis. In particle language this is the $U(1)$ gauge symmetry associated with particle number conservation of the Hamiltonian (11.1). Off-diagonal long range order (or, equivalently, BEC) implies that this symmetry is spontaneously broken in the state under consideration.[1] It is notoriously difficult to prove such symmetry breaking for systems with a continuous symmetry. One of the few available techniques is that of *reflection positivity* (and the closely related property of *Gaussian domination*) and fortunately it can be applied to our system. For this, however, the hard core and half-filling conditions are essential because they imply a particle-hole symmetry that is crucial for the proofs to work. Naturally, BEC is expected to occur at other fillings, but no one has so far found a way to prove condensation (or, equivalently, long-range order in an antiferromagnet with continuous symmetry) without using reflection positivity and infrared bounds, and these require the additional symmetry.

Reflection positivity was first formulated by K. Osterwalder and R. Schrader [OS] in the context of relativistic quantum field theory. Later, J. Fröhlich, B. Simon and T. Spencer used the concept to prove the existence of a phase transition for a classical spin model with a continuous symmetry [FSS], and E. Lieb and J. Fröhlich [FL] as well as F. Dyson, E. Lieb and B. Simon [DLS] applied it for the analysis of quantum spin systems. The proof of off-diagonal long range order for the Hamiltonian (11.2) (for small λ) given here is based on appropriate modifications of the arguments in [DLS].

---

[1] See the discussion at the end of Appendix D.

11.2   Reflection Positivity

In the present context reflection positivity means the following. We divide the torus Λ into two congruent parts, Λ_L and Λ_R, by cutting it with a hyperplane orthogonal to one of the d directions. (For this we assume that the side length of Λ is even.) This induces a factorization of the Hilbert space, $\mathcal{H} = \mathcal{H}_L \otimes \mathcal{H}_R$, with

$$\mathcal{H}_{L,R} = \bigotimes_{\mathbf{x} \in \Lambda_{L,R}} \mathbb{C}^2.$$

There is a natural identification between a site $\mathbf{x} \in \Lambda_L$ and its mirror image $\vartheta\mathbf{x} \in \Lambda_R$. If F is an operator on $\mathcal{H} = \mathcal{H}_L$ we define its reflection θF as an operator on $\mathcal{H}_R$ in the following way. If $F = F_\mathbf{x}$ operates non-trivially only on one site, $\mathbf{x} \in \Lambda_L$, we define $\theta F = V F_{\vartheta\mathbf{x}} V^\dagger$ where V denotes the unitary particle-hole transformation or, in the spin language, rotation by π around the 1-axis. This definition extends in an obvious way to products of operators on single sites and then, by linearity, to arbitrary operators on $\mathcal{H}_L$. Reflection positivity of a state $\langle \cdot \rangle$ means that

$$\langle F\theta\overline{F} \rangle \geq 0 \tag{11.3}$$

for any F operating on $\mathcal{H}_L$. Here $\overline{F}$ is the complex conjugate of the operator F in the matrix representation defined above, i.e., defined by the basis where the operators $S_\mathbf{x}^3$ are diagonal.

We now show that reflection positivity holds for any thermal equilibrium state of our Hamiltonian. We can write the Hamiltonian (11.2) as

$$H = H_L + H_R - \tfrac{1}{2} \sum_{\langle \mathbf{xy} \rangle \in M} (S_\mathbf{x}^+ S_\mathbf{y}^- + S_\mathbf{x}^- S_\mathbf{y}^+), \tag{11.4}$$

where H_L and H_R act non-trivially only on $\mathcal{H}_L$ and $\mathcal{H}_R$, respectively. Here, M denotes the set of bonds going from the left sublattice to the right sublattice. (Because of the periodic boundary condition these include the bonds that connect the right boundary with the left boundary.) Note that $H_R = \theta H_L$, and

$$\sum_{\langle \mathbf{xy} \rangle \in M} (S_\mathbf{x}^+ S_\mathbf{y}^- + S_\mathbf{x}^- S_\mathbf{y}^+) = \sum_{\langle \mathbf{xy} \rangle \in M} (S_\mathbf{x}^+ \theta S_\mathbf{x}^+ + S_\mathbf{x}^- \theta S_\mathbf{x}^-).$$

For these properties it is essential that we included the unitary particle-hole transformation V in the definition of the reflection θ. For reflection positivity it is also important that all operators appearing in H (11.4) have a *real* matrix representation. Moreover, the minus sign in (11.4) is essential.

Using the Trotter product formula, we have

$$\text{tr } F\theta\overline{F}e^{-\beta H} = \lim_{n\to\infty} \text{tr } F\theta\overline{F}\, \mathcal{Z}_n$$

with

$$\mathcal{Z}_n = \left[e^{-\frac{1}{n}\beta H_L}\theta e^{-\frac{1}{n}\beta H_L} \prod_{\langle \mathbf{xy}\rangle \in M} \left(1 + \frac{\beta}{2n}[S_{\mathbf{x}}^+\theta S_{\mathbf{x}}^+ + S_{\mathbf{x}}^-\theta S_{\mathbf{x}}^-)] \right) \right]^n. \quad (11.5)$$

Observe that $\mathcal{Z}_n$ is a sum of terms of the form

$$\prod_i A_i\theta A_i, \quad (11.6)$$

with A_i given by either $e^{-\frac{1}{n}\beta H_L}$ or $\sqrt{\frac{\beta}{2n}}S_{\mathbf{x}}^+$ or $\sqrt{\frac{\beta}{2n}}S_{\mathbf{x}}^-$. All the A_i are real matrices, and therefore

$$\text{tr}_{\mathcal{H}}\, F\theta\overline{F}\prod_i A_i\theta A_i = \text{tr}_{\mathcal{H}}\, F\prod_i A_i\, \theta\left[\overline{F}\prod_j A_j\right] = \left|\text{tr}_{\mathcal{H}_L}\, F\prod_i A_i\right|^2 \geq 0. \quad (11.7)$$

Hence $\text{tr } F\theta\overline{F}\, \mathcal{Z}_n$ is a sum of non-negative terms and therefore non-negative. This proves our assertion.

11.3 Proof of BEC for Small λ and T

The main tool in our proof of BEC are *infrared bounds*. More precisely, for $\mathbf{p} \in \Lambda^*$ (the dual lattice of Λ), let $\widetilde{S}_{\mathbf{p}}^{\#} = |\Lambda|^{-1/2}\sum_{\mathbf{x}} S_{\mathbf{x}}^{\#}\exp(i\mathbf{p}\cdot\mathbf{x})$ denote the Fourier transform of the spin operators. We claim that

$$(\widetilde{S}_{\mathbf{p}}^1, \widetilde{S}_{-\mathbf{p}}^1) \leq \frac{T}{2E_{\mathbf{p}}}, \quad (11.8)$$

with $E_{\mathbf{p}} = \sum_{i=1}^d (1 - \cos(p_i))$. Here, p_i denotes the components of $\mathbf{p}$, and $(\,,\,)$ denotes the Duhamel two point function at temperature T, defined by

$$(A, B) = \int_0^1 \text{tr}\left(Ae^{-s\beta H}Be^{-(1-s)\beta H}\right) ds/\text{tr}\, e^{-\beta H} \quad (11.9)$$

for any pair of operators A and B. Because of invariance under rotations around the S^3 axis, (11.8) is equally true with S^1 replaced by S^2, of course.

The crucial lemma (*Gaussian domination*) is the following. Define, for a complex valued function h on the bonds $\langle \mathbf{xy}\rangle$ in Λ,

$$Z(h) = \text{tr}\exp\left[-\beta K(h)\right], \quad (11.10)$$

with $K(h)$ the modified Hamiltonian

$$K(h) = \frac{1}{4} \sum_{\langle \mathbf{xy} \rangle} \left(\left(S_{\mathbf{x}}^+ - S_{\mathbf{y}}^- - h_{\mathbf{xy}} \right)^2 + \left(S_{\mathbf{x}}^- - S_{\mathbf{y}}^+ - \overline{h_{\mathbf{xy}}} \right)^2 \right) + \lambda \sum_{\mathbf{x}} \left[\tfrac{1}{2} + (-1)^{\mathbf{x}} S_{\mathbf{x}}^3 \right].$$

$$(11.11)$$

Note that for $h \equiv 0$, $K(h)$ agrees with the Hamiltonian H, because $(S^\pm)^2 = 0$. We claim that, for any *real valued* h,

$$Z(h) \leq Z(0). \tag{11.12}$$

The infrared bound then follows from $d^2 Z(\varepsilon h)/d\varepsilon^2|_{\varepsilon=0} \leq 0$, taking $h_{\mathbf{xy}} = \exp(\mathrm{i}\mathbf{p} \cdot \mathbf{x}) - \exp(\mathrm{i}\mathbf{p} \cdot \mathbf{y})$. This is not a real function, though, but the negativity of the (real!) quadratic form $d^2 Z(\varepsilon h)/d\varepsilon^2|_{\varepsilon=0}$ for real h implies negativity also for complex-valued h.

The proof of (11.12) is very similar to the proof of the reflection positivity property (11.3) given above. It follows along the same lines as in [DLS], but we repeat it here for convenience of the reader.

The intuition behind (11.12) is the following. First, in maximizing $Z(h)$ one can restrict to gradients, i.e., $h_{\mathbf{xy}} = \hat{h}_{\mathbf{x}} - \hat{h}_{\mathbf{y}}$ for some function $\hat{h}_{\mathbf{x}}$ on Λ. (This follows from stationarity of $Z(h)$ at a maximizer $h_{\max}$.) Reflection positivity implies that $\langle A\theta \overline{B} \rangle$ defines a scalar product on operators on $\mathcal{H}_L$, and hence there is a corresponding Schwarz inequality. Moreover, since reflection positivity holds for reflections across *any* hyperplane, one arrives at the so-called *chessboard inequality*, which is simply a version of Schwarz's inequality for multiple reflections across different hyperplanes. Such a chessboard estimate implies that in order to maximize $Z(h)$ it is best to choose the function $\hat{h}_{\mathbf{x}}$ to be constant. In the case of classical spin systems [FSS], this intuition can be turned into a complete proof of (11.12). Because of non-commutativity of $K(h)$ with $K(0) = H$, this is not possible in the quantum case. However, one can proceed by using the Trotter formula as follows.

Let $h_{\max}$ be a function that maximizes $Z(h)$ for real valued h. If there is more than one maximizer, we choose $h_{\max}$ to be one that vanishes on the largest number of bonds. We then have to show that actually $h_{\max} \equiv 0$. If $h_{\max} \neq 0$, we draw a hyperplane such that $h_{\mathbf{xy}} \neq 0$ for at least one pair $\langle \mathbf{xy} \rangle$ crossing the plane. We can again write

$$K(h) = K_L(h) + K_R(h) + \frac{1}{4} \sum_{\langle \mathbf{xy} \rangle \in M} \left((S_{\mathbf{x}}^+ - S_{\mathbf{y}}^- - h_{\mathbf{xy}})^2 + (S_{\mathbf{x}}^- - S_{\mathbf{y}}^+ - h_{\mathbf{xy}})^2 \right).$$

$$(11.13)$$

Using the Trotter formula, we have $Z(h) = \lim_{n\to\infty} \alpha_n$, with

$$\alpha_n = \text{tr} \left[e^{-\beta K_L/n} e^{-\beta K_R/n} \prod_{\langle \mathbf{xy} \rangle \in M} e^{-\beta(S_\mathbf{x}^+ - S_\mathbf{y}^- - h_{\mathbf{xy}})^2/4n} e^{-\beta(S_\mathbf{x}^- - S_\mathbf{y}^+ - h_{\mathbf{xy}})^2/4n} \right]^n.$$

$$(11.14)$$

For any matrix, we can write

$$e^{-D^2} = (4\pi)^{-1/2} \int_{\mathbb{R}} dk\, e^{ikD} e^{-k^2/4}.$$

$$(11.15)$$

If we apply this to the last two factors in (11.14), and note that $S_\mathbf{y}^- = \theta S_\mathbf{x}^+$ if $\langle \mathbf{xy} \rangle \in M$. Denoting by $\mathbf{x}_1, \ldots, \mathbf{x}_l$ the points on the left side of the bonds in M, we have that

$$\alpha_n = (4\pi)^{-nl} \int_{R^{2nl}} d^{2nl}k\, \text{tr} \left[e^{-\beta K_L/n} e^{-\beta K_R/n} e^{ik_1(S_{\mathbf{x}_1}^+ - \theta S_{\mathbf{x}_1}^+)\beta^{1/2}/2n^{1/2}} \cdots \right]$$

$$\times e^{-k^2/4} e^{-ik_1 h_{\mathbf{x}_1} \vartheta \mathbf{x}_1 \beta^{1/2}/2n^{1/2}\cdots}.$$

$$(11.16)$$

Here we denote $k^2 = \sum k_i^2$ for short. Since matrices on the right of M commute with matrices on the left, and since all matrices in question are *real*, we see that the trace in the integrand above can be written as

$$\text{tr} \left[e^{-\beta K_L/n} e^{ik_1 S_{\mathbf{x}_1}^+ \beta^{1/2}/2n^{1/2}} \cdots \right] \overline{\text{tr} \left[e^{-\beta K_R/n} e^{ik_1 \theta S_{\mathbf{x}_1}^+ \beta^{1/2}/2n^{1/2}} \cdots \right]}.$$

$$(11.17)$$

Using the Schwarz inequality for the k integration, and 'undoing' the above step, we see that

$$|\alpha_n|^2 \leq \left((4\pi)^{-nl} \int_{R^{2nl}} d^{2nl}k\, e^{-k^2/4} \right.$$

$$\times \text{tr} \left[e^{-\beta K_L/n} e^{-\beta \theta K_L/n} e^{ik_1(S_{\mathbf{x}_1}^+ - \theta S_{\mathbf{x}_1}^+)\beta^{1/2}/2n^{1/2}} \cdots \right] \right)$$

$$\times \left((4\pi)^{-nl} \int_{R^{2nl}} d^{2nl}k\, e^{-k^2/4} \right.$$

$$\left. \times \text{tr} \left[e^{-\beta \theta K_R/n} e^{-\beta K_R/n} e^{ik_1(S_{\mathbf{x}_1}^+ - \theta S_{\mathbf{x}_1}^+)\beta^{1/2}/2n^{1/2}} \cdots \right] \right). \quad (11.18)$$

In terms of the partition function $Z(h)$, this means that

$$|Z(h_{\text{max}})|^2 \leq Z(h^{(1)}) Z(h^{(2)}),$$

$$(11.19)$$

where $h^{(1)}$ and $h^{(2)})$ are obtained from h_{max} by reflection across M in the following way:

$$h_{\mathbf{xy}}^{(1)} = \begin{cases} h_{\mathbf{xy}} & \text{if } \mathbf{x}, \mathbf{y} \in \Lambda_L \\ h_{\vartheta \mathbf{x} \vartheta \mathbf{y}} & \text{if } \mathbf{x}, \mathbf{y} \in \Lambda_R \\ 0 & \text{if } \langle \mathbf{xy} \rangle \in M \end{cases}$$

$$(11.20)$$

and $h^{(2)}$ is given by the same expression, interchanging L and R. Therefore also $h^{(1)}$ and $h^{(2)}$ must be maximizers of $Z(h)$. However, one of them will contain strictly more zeros than $h_{\max}$, since $h_{\max}$ does not vanish identically for bonds crossing M. This contradicts our assumption that $h_{\max}$ contains the maximal number of zeros among all maximizers of $Z(h)$. Hence $h_{\max} \equiv 0$ identically. This completes the proof of (11.12).

The next step is to transfer the upper bound on the Duhamel two point function (11.8) into an upper bound on the thermal expectation value. This involves convexity arguments and estimations of double commutators like in Section 3 in [DLS]. For this purpose, we have to evaluate the double commutators

$$[\widetilde{S}^1_{\mathbf{p}}, [H, \widetilde{S}^1_{-\mathbf{p}}]] + [\widetilde{S}^2_{\mathbf{p}}, [H, \widetilde{S}^2_{-\mathbf{p}}]] = -\frac{2}{|\Lambda|}\Big(H - \tfrac{1}{2}\lambda|\Lambda| + 2\sum_{\langle \mathbf{xy}\rangle} S^3_{\mathbf{x}}S^3_{\mathbf{y}}\cos\mathbf{p}\cdot(\mathbf{x}-\mathbf{y})\Big).$$

$$(11.21)$$

Let $C_{\mathbf{p}}$ denote the expectation value of this last expression,

$$C_{\mathbf{p}} = \langle [\widetilde{S}^1_{\mathbf{p}}, [H, \widetilde{S}^1_{-\mathbf{p}}]] + [\widetilde{S}^2_{\mathbf{p}}, [H, \widetilde{S}^2_{-\mathbf{p}}]]\rangle \geq 0.$$

The positivity of $C_{\mathbf{p}}$ can be seen from an eigenfunction-expansion of the trace. From [DLS, Corollary 3.2 and Theorem 3.2] and (11.8) we infer that

$$\langle \widetilde{S}^1_{\mathbf{p}}\widetilde{S}^1_{-\mathbf{p}} + \widetilde{S}^2_{\mathbf{p}}\widetilde{S}^2_{-\mathbf{p}}\rangle \leq \frac{1}{2}\sqrt{\frac{C_{\mathbf{p}}}{E_{\mathbf{p}}}}\coth\sqrt{\beta^2 C_{\mathbf{p}}E_{\mathbf{p}}/4}. \qquad (11.22)$$

Using $\coth x \leq 1 + 1/x$ and Schwarz's inequality, we obtain for the sum over all $\mathbf{p} \neq \mathbf{0}$,

$$\sum_{\mathbf{p}\neq 0}\langle \widetilde{S}^1_{\mathbf{p}}\widetilde{S}^1_{-\mathbf{p}} + \widetilde{S}^2_{\mathbf{p}}\widetilde{S}^2_{-\mathbf{p}}\rangle \leq \frac{1}{\beta}\sum_{\mathbf{p}\neq 0}\frac{1}{E_{\mathbf{p}}} + \frac{1}{2}\Big(\sum_{\mathbf{p}\neq 0}\frac{1}{E_{\mathbf{p}}}\Big)^{1/2}\Big(\sum_{\mathbf{p}\neq 0}C_{\mathbf{p}}\Big)^{1/2}. \qquad (11.23)$$

We have $\sum_{\mathbf{p}\in\Lambda^*} C_{\mathbf{p}} = -2\langle H\rangle + \lambda|\Lambda|$, which can be bounded from above using the following lower bound on the Hamiltonian:

$$H \geq -\frac{|\Lambda|}{4}\big[d(d+1) + 4\lambda^2\big]^{1/2} + \tfrac{1}{2}\lambda|\Lambda|. \qquad (11.24)$$

This inequality follows from the fact that the lowest eigenvalue of

$$-\frac{1}{2}S^1_{\mathbf{x}}\sum_{i=1}^{2d} S^1_{\mathbf{y}_i} - \frac{1}{2}S^2_{\mathbf{x}}\sum_{i=1}^{2d} S^2_{\mathbf{y}_i} + \lambda S^3_{\mathbf{x}} \qquad (11.25)$$

is given by $-\frac{1}{4}[d(d+1) + 4\lambda^2]^{1/2}$. This can be shown exactly in the same way as [DLS, Theorem C.1]. Since the Hamiltonian H can be written as a sum of terms like (11.25), with $\mathbf{y}_i$ the nearest neighbors of $\mathbf{x}$, we get from this fact the lower bound (11.24).

With the aid of the sum rule

$$\sum_{\mathbf{p}\in\Lambda^*} \langle \tilde{S}^1_{\mathbf{p}}\tilde{S}^1_{-\mathbf{p}} + \tilde{S}^2_{\mathbf{p}}\tilde{S}^2_{-\mathbf{p}}\rangle = \frac{|\Lambda|}{2}$$

(which follows from $(S^1)^2 = (S^2)^2 = 1/4$), we obtain from (11.23) and (11.24) the following lower bound in the thermodynamic limit:

$$\lim_{\Lambda\to\infty} \frac{1}{|\Lambda|}\langle \tilde{S}^1_0\tilde{S}^1_0 + \tilde{S}^2_0\tilde{S}^2_0\rangle$$

$$\geq \frac{1}{2} - \frac{1}{2}\left(\tfrac{1}{2}\left[d(d+1) + 4\lambda^2\right]^{1/2} c_d\right)^{1/2} - \frac{1}{\beta}c_d, \qquad (11.26)$$

with c_d given by

$$c_d = \frac{1}{(2\pi)^d}\int_{[-\pi,\pi]^d} d\mathbf{p}\,\frac{1}{E_{\mathbf{p}}}. \qquad (11.27)$$

This is our final result. Note that c_d is finite for $d \geq 3$. Hence the right side of (11.26) is positive, for large enough β, as long as

$$\lambda^2 < \frac{1}{c_d^2} - \frac{d(d+1)}{4}.$$

In $d = 3$, $c_3 \approx 0.505$ [DLS], and hence this condition is fulfilled for $\lambda \lesssim 0.960$. In [DLS] it was also shown that dc_d is monotone decreasing in d, which implies a similar result for all $d > 3$.

The connection with BEC is as follows. Since H is real, also $\gamma(\mathbf{x}, \mathbf{y})$ is real and we have

$$\gamma(\mathbf{x}, \mathbf{y}) = \langle S^+_\mathbf{x} S^-_\mathbf{y}\rangle = \langle S^1_\mathbf{x}S^1_\mathbf{y} + S^2_\mathbf{x}S^2_\mathbf{y}\rangle.$$

Hence, if $\varphi_0 = |\Lambda|^{-1/2}$ denotes the constant function,

$$\langle\varphi_0|\gamma|\varphi_0\rangle = \langle \tilde{S}^1_0\tilde{S}^1_0 + \tilde{S}^2_0\tilde{S}^2_0\rangle,$$

and thus the bound (11.26) implies that the largest eigenvalue of $\gamma(\mathbf{x}, \mathbf{y})$ is bounded from below by the right side of (11.26). In addition one can show that the infrared bounds imply that there is at most *one* large eigenvalue (of the order $|\Lambda|$), and that the corresponding eigenvector (the 'condensate wave function') is strictly constant in the thermodynamic limit [ALSSY]. The constancy of the condensate wave function is surprising and is not expected to hold for densities different from $\frac{1}{2}$, where particle-hole symmetry is absent. In contrast to the condensate wave function the particle density shows the staggering of the periodic potential [ALSSY, Thm. 3]. It also contrasts with the situation for zero interparticle interaction, as discussed at the end of this chapter.

In the BEC phase there is *no gap* for adding particles beyond half filling (in the thermodynamic limit): The ground state energy, E_k, for $\frac{1}{2}|\Lambda| + k$ particles satisfies

$$0 \le E_k - E_0 \le \frac{(\text{const.})}{|\Lambda|} \tag{11.28}$$

(with a constant that depends on k but not on $|\Lambda|$.) The proof of (11.28) is by a variational calculation, with a trial state of the form $(\widetilde{S}_0^+)^k|0\rangle$, where $|0\rangle$ denotes the absolute ground state, i.e., the ground state for half filling. (This is the unique ground state of the Hamiltonian, as can be shown using reflection positivity. See Appendix A in [ALSSY].) Also, in the thermodynamic limit, the energy per site for a given density, $e(\varrho)$, satisfies

$$e(\varrho) - e(\tfrac{1}{2}) \le \text{const.} \, (\varrho - \tfrac{1}{2})^2. \tag{11.29}$$

Thus there is no cusp at $\varrho = 1/2$. To show this, one takes a trial state of the form

$$|\psi_\mathbf{y}\rangle = e^{i\varepsilon \sum_\mathbf{x} S_\mathbf{x}^2}(S_\mathbf{y}^1 + \tfrac{1}{2})|0\rangle. \tag{11.30}$$

The motivation is the following: we take the ground state and first project onto a given direction of S^1 on some site $\mathbf{y}$. If there is long-range order, this should imply that essentially all the spins point in this direction now. Then we rotate slightly around the S^2-axis. The particle number should then go up by $\varepsilon|\Lambda|$, but the energy only by $\varepsilon^2|\Lambda|$. We refer to [ALSSY, Sect. IV] for the details.

The absence of a gap in the case of BEC is not surprising, since a gap is characteristic for a Mott insulator state. We show the occurrence of a gap, for large enough λ, in the next section.

11.4   Absence of BEC and Mott Insulator Phase

The main results of this section are the following: If either

- $\lambda \ge 0$ and $T > d/(2\ln 2)$, or

- $T \ge 0$ and $\lambda \ge 0$ such that $\lambda + |e(\lambda)| > d$, with $e(\lambda) = $ ground state energy per site, then there is exponential decay of correlations:

$$\gamma(\mathbf{x}, \mathbf{y}) \leq (\text{const.}) \exp(-\kappa |\mathbf{x} - \mathbf{y}|) \tag{11.31}$$

with $\kappa > 0$. Moreover, for $T = 0$, the ground state energy in a sector of fixed particle number $N = \frac{1}{2}|\Lambda| + k$, denoted by E_k, satisfies

$$E_k + E_{-k} - 2E_0 \geq (\lambda + |e(\lambda)| - d)|k|. \tag{11.32}$$

I.e, for large enough λ the chemical potential has a jump at half filling.

The derivation of these two properties is based on a path integral representation of the equilibrium state at temperature T, and of the ground state which is obtained in the limit $T \to \infty$. density matrix. The analysis starts from the observation that the density operator $e^{-\beta H}$ has non-negative matrix elements in the basis in which $\{S^3_\mathbf{x}\}$ are diagonal, i.e. of states with specified particle occupation numbers. It is convenient to focus on the dynamics of the 'quasi-particles' which are defined so that the presence of one at a site $\mathbf{x}$ signifies a deviation there from the occupation state which minimizes the potential-energy. Since the Hamiltonian is $H = H_0 + \lambda W$, with H_0 the hopping term in (11.2) and W the staggered field, we define the quasi-particle number operators $n_\mathbf{x}$ as:

$$n_\mathbf{x} = \tfrac{1}{2} + (-1)^\mathbf{x} S^3_\mathbf{x} = \begin{cases} a^\dagger_\mathbf{x} a_\mathbf{x} & \text{for } \mathbf{x} \text{ even} \\ 1 - a^\dagger_\mathbf{x} a_\mathbf{x} & \text{for } \mathbf{x} \text{ odd} \end{cases}. \tag{11.33}$$

Thus $n_\mathbf{x} = 1$ means presence of a particle if $\mathbf{x}$ is on the A sublattice (potential maximum) and absence if $\mathbf{x}$ is on the B sublattice (potential minimum).

The collection of the joint eigenstates of the occupation numbers, $\{|\{n_\mathbf{x}\}\rangle\}$, provides a convenient basis for the Hilbert space. The functional integral representation of $\langle \{n_\mathbf{x}\}| e^{-\beta(H_0 + \lambda W)} |\{n_\mathbf{x}\}\rangle$ involves an integral over configurations of quasi-particle loops in a *space* $\times$ *time* for which the (imaginary) 'time' corresponds to a variable with period β. The fact that the integral is over a positive measure facilitates the applicability of statistical-mechanics intuition and tools. One finds that the quasi-particles are suppressed by the potential energy, but favored by the entropy, which enters this picture due to the presence of the hopping term in H. At large λ, the potential suppression causes localization: long 'quasi-particle' loops are rare, and the amplitude for long paths decays exponentially in the distance, both for paths which may occur spontaneously and for paths whose presence is forced through the insertion of sources, i.e., particle creation and annihilation operators. Localization is also caused by high temperature, since the requirement of periodicity implies that at any site which participates in a loop there should be at least two jumps during the short 'time' interval $[0, \beta)$ and the amplitude for even a single jump is small, of order β.

The path integral described above is obtained through the Dyson expansion

$$e^{t(A+B)} = e^{tA} \sum_{m \geq 0} \int_{0 \leq t_1 \leq t_2 \leq \cdots \leq t_m \leq t} B(t_m) \cdots B(t_1) dt_1 \cdots dt_m \tag{11.34}$$

for any matrices A and B and $t > 0$, with $B(t) = e^{-tA} B e^{tA}$. (The $m = 0$ term in the sum is interpreted here as 1.)

In evaluating the matrix elements of $e^{-\beta H} = e^{-\beta(H_0 + \lambda W)}$, in the basis $\{|\{n_\mathbf{x}\}\rangle\}$, we note that W is diagonal and $\langle\{n_\mathbf{x}\}|H_0|\{n'_\mathbf{x}\}\rangle$ are non-zero only if the configurations $\{n_\mathbf{x}\}$ and $\{n'_\mathbf{x}\}$ differ at exactly one nearest neighbor pair of sites where the change corresponds to either a creation of a pair of quasi-particles or the annihilation of such a pair. I.e., the matrix elements is zero unless $n_\mathbf{x} = n'_\mathbf{x}$ for all $\mathbf{x}$ except for a nearest neighbor pair $\langle \mathbf{xy} \rangle$, where $n_\mathbf{x} = n_\mathbf{y}$, $n'_\mathbf{x} = n'_\mathbf{y}$, and $n_\mathbf{x} + n'_\mathbf{x} = 1$. In this case, the matrix element equals $-1/2$.

Introducing intermediate states, the partition function can thus be written as follows:

$$\text{tr } e^{-\beta H} = \sum_{m=0}^{\infty} \int_{0 \leq t_1 \leq t_2 \leq \cdots \leq t_m \leq \beta} \sum_{|\{n_\mathbf{x}^{(i)}\}\rangle, 1 \leq i \leq m}$$

$$\times \exp\left(-\lambda \sum_{i=1}^{m}(t_i - t_{i-1}) \sum_{\mathbf{x}} n_\mathbf{x}^{(i)}\right) dt_1 \cdots dt_m$$

$$\times (-1)^m \langle\{n_\mathbf{x}^{(1)}\}|H_0|\{n_\mathbf{x}^{(m)}\}\rangle\langle\{n_\mathbf{x}^{(m)}\}|H_0|\{n_\mathbf{x}^{(m-1)}\}\rangle$$

$$\times \langle\{n_\mathbf{x}^{(m-1)}\}|H_0|\{n_\mathbf{x}^{(m-2)}\}\rangle \cdots \langle\{n_\mathbf{x}^{(2)}\}|H_0||\{n_\mathbf{x}^{(1)}\}\rangle \tag{11.35}$$

with the interpretation $t_0 = t_m - \beta$. Note that the factor in the last two lines of (11.35) equals $(1/2)^m$ if adjacent elements in the sequence of configurations $\{n_\mathbf{x}^{(i)}\}$ differ by exactly one quasi-particle pair, otherwise it is zero.

Expansions of this type are explained more fully in [AN]. A compact way of writing (11.35) is:

$$\text{tr } e^{-\beta H} = \int v(d\omega) e^{-\lambda|\omega|}. \tag{11.36}$$

Here the 'path' ω stands for a set of disjoint oriented loops in the 'space-time' $\Lambda \times [0, \beta]$, with periodic boundary conditions in 'time'. Each ω is parametrized by a number of jumps, m, jumping times $0 \leq t_1 \leq t_2 \leq \cdots \leq t_m \leq \beta$, and a sequence of configurations $\{n_\mathbf{x}^{(i)}\}$, which is determined by the initial configuration $\{n_\mathbf{x}^{(1)}\}$ plus a sequence of 'rungs' connecting nearest neighbor sites, depicting the creation or annihilation of a pair of neighboring quasi-particles (see Fig. 11.2). As in Feynman's picture of QED, it is convenient to regard such an event as a jump of the quasi-particle, at which its time-orientation is also reversed. The length of ω, denoted by $|\omega|$, is the sum of the vertical lengths of the loops. The measure $v(d\omega)$ is determined by (11.35); namely, for a given sequence of configurations $\{n_\mathbf{x}^{(i)}\}$, $1 \leq i \leq m$, the integration takes places over the times of the jumps, with a measure $(1/2)^m dt_1 \cdots dt_m$.

One may note that the measure $v(d\omega)$ corresponds to a Poisson process of random configurations of oriented 'rungs', linking neighboring sites at random times, and signifying either the creation or the annihilation of a pair of quasi-particles. The matrix element $\langle\{n_\mathbf{x}\}|e^{-\beta H}|\{n_\mathbf{x}'\}\rangle$ gets no contribution from rung configurations that are inconsistent, either internally or with the boundary conditions corresponding to the specified state vectors. A consistent configuration yields a family of non-overlapping loops which describe the motion of the quasi-particles in the 'space-time' $\Lambda \times [0, \beta)$. Each such configuration contributes with weight $e^{-\lambda|\omega|}$ to the above matrix element (another positive factor was absorbed in the measure $v(d\omega)$). One may note that long paths are suppressed in the integral (11.38) at a rate which increases with λ.

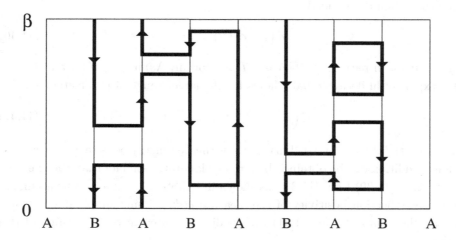

Figure 11.2: Loop gas describing paths of quasi-particles for particle number $N = |\Lambda|/2 - 1$. A line on an A site means presence of a particle, while on a B site it means absence. The horizontal rungs correspond to hopping of a particle.

Likewise, for $\mathbf{x} \neq \mathbf{y}$, we can write

$$\operatorname{tr} a_{\mathbf{x}}^\dagger a_{\mathbf{y}} e^{-\beta H} = \int_{\mathcal{A}^{(\mathbf{x},\mathbf{y})}} v(d\omega) e^{-\lambda|\omega|}, \tag{11.37}$$

where $\mathcal{A}^{(\mathbf{x},\mathbf{y})}$ denotes the set of all loops that, besides disjoint closed loops, contain one curve which avoids all the loops and connects $\mathbf{x}$ and $\mathbf{y}$ at time zero. The one-particle density matrix can thus be written

$$\gamma(\mathbf{x}, \mathbf{y}) = \frac{\int_{\mathcal{A}^{(\mathbf{x},\mathbf{y})}} v(d\omega) e^{-\lambda|\omega|}}{\int v(d\omega) e^{-\lambda|\omega|}}. \tag{11.38}$$

For an upper bound, we can drop the condition in the numerator that the loops and the curve from $\mathbf{x}$ to $\mathbf{y}$ do not intersect. The resulting measure space is simply a Cartesian product of the measure space appearing in the denominator and the space of all curves, ζ, connecting $\mathbf{x}$ and $\mathbf{y}$, both at time 0. Denoting the latter by $\mathcal{B}(\mathbf{x}, \mathbf{y})$, we thus get the upper bound

$$\gamma(\mathbf{x}, \mathbf{y}) \leq \int_{\mathcal{B}(\mathbf{x},\mathbf{y})} v(d\zeta) e^{-\lambda|\zeta|}. \tag{11.39}$$

The integral over paths is convergent if either λ or T is small enough, and away from the convergence threshold the resulting amplitude decays exponentially. A natural random walk estimate, see [ALSSY, Lemma 4], leads to the claimed exponential bound provided

$$d\left(1 - e^{-\beta\lambda}\right) < \lambda. \tag{11.40}$$

This includes, in particular, the cases $T > d$ for any λ, and $\lambda > d$ for any T.

Exponential decay actually holds for the larger range of parameters where

$$d\left(1 - e^{-\beta(\lambda-f)}\right) < \lambda - f, \tag{11.41}$$

where $f = f(\beta, \lambda) = -(\beta|\Lambda|)^{-1} \ln \operatorname{tr} e^{-\beta H}$ is the free energy per site. Note that $f < 0$. This condition can be obtained by a more elaborate estimate than the one used in obtaining (11.39) from (11.38), as shown in [ALSSY, Lemma 3]. The argument there uses reflection positivity of the measure $v(d\omega)$. Using simple bounds on f one can then obtain from (11.41) the conditions stated in the beginning of this section.

The proof of the energy gap is based on an estimate for the ratio $\frac{\operatorname{tr} \mathcal{P}_k e^{-\beta H}}{\operatorname{tr} \mathcal{P}_0 e^{-\beta H}}$ where $\mathcal{P}_k$ projects onto states in Fock space with particle number $N = \frac{1}{2}|\Lambda| + k$, expressing numerator and denominator in terms of path integrals. The integral for the numerator is over configurations ω with a non-trivial winding number k. Each such configuration includes a collection of 'non-contractible' loops with total length at least $\beta|k|$. An estimate of the relative weight of such loops yields the bound

$$\frac{\operatorname{tr} \mathcal{P}_k e^{-\beta H}}{\operatorname{tr} \mathcal{P}_0 e^{-\beta H}} \leq (\text{const.})(|\Lambda|/|k|)^{|k|} \left(e^{1-(\text{const.})\beta}\right)^{|k|} \tag{11.42}$$

which gives for $\beta \to \infty$

$$E_k - E_0 \geq \text{const.} \, |k| \tag{11.43}$$

independently of $|\Lambda|$. We refer to [ALSSY] for details.

11.5 The Non-Interacting Gas

The interparticle interaction is essential for the existence of a Mott insulator phase for large λ. In case of absence of the hard-core interaction, there is BEC for any density and any λ at low enough temperature (for $d \geq 3$). To see this, we have to calculate the spectrum of the one-particle Hamiltonian $-\frac{1}{2}\Delta + V(\mathbf{x})$, where Δ denotes the discrete Laplacian and $V(\mathbf{x}) = \lambda(-1)^{\mathbf{x}}$. The spectrum can be easily obtained by noting that V anticommutes with the off-diagonal part of the Laplacian, i.e., $\{V, \Delta + 2d\} = 0$. Hence

$$\left(-\tfrac{1}{2}\Delta - d + V(\mathbf{x})\right)^2 = \left(-\tfrac{1}{2}\Delta - d\right)^2 + \lambda^2, \tag{11.44}$$

so the spectrum is given by

$$d \pm \sqrt{\left(\textstyle\sum_i \cos p_i\right)^2 + \lambda^2}, \tag{11.45}$$

where $\mathbf{p} \in \Lambda^*$. In particular, $E(\mathbf{p}) - E(0) \sim \frac{1}{2}d(d^2 + \lambda^2)^{-1/2}|\mathbf{p}|^2$ for small $|\mathbf{p}|$, and hence there is BEC for low enough temperature. Note that the condensate wave function is of course *not* constant in this case, but rather given by the eigenfunction corresponding to the lowest eigenvalue of $-\frac{1}{2}\Delta + \lambda(-1)^{\mathbf{x}}$.

11.6 Conclusion

In this chapter a lattice model is studied, which is similar to the usual Bose-Hubbard model and which describes the transition between Bose-Einstein condensation and a Mott insulator state as the strength λ of an optical lattice potential is increased. While the model is not soluble in the usual sense, it is possible to prove rigorously all the essential features that are observed experimentally. These include the existence of BEC for small λ and its suppression for large λ, which is a localization phenomenon depending heavily on the fact that the Bose particles interact with each other. The Mott insulator regime is characterized by a gap in the chemical potential, which does not exist in the BEC phase and for which the interaction is also essential. It is possible to derive bounds on the critical λ as a function of temperature.

Appendix A

Elements of Bogoliubov Theory

This appendix is Section 2 of the 1965 review article [L3], with minor modifications. Unlike the rest of these notes, this appendix is not mathematically rigorous, but it is important because it shows what is widely believed to be the case about the low end of the spectrum of the Hamiltonian H_N — even if it cannot be proved. It is also a heuristic guide to thinking about Bose gases.

In the forty years since the appearance of the review article [L3] the Bogoliubov model has been extensively studied; an account of many of these developments can be found in the comprehensive article [ZB].

Almost all work in the field of interacting Bose gases has its genesis in Bogoliubov's 1947 paper [Bo] and we shall begin with a brief summary of its main features. His theory is important because it set us on the right track. But of equal importance is understanding, with a little hindsight, the failures of the theory, for its success led to a regrettable tendency to take its predictions as gospel. Moreover, like many great approximation schemes in mathematical physics, the first order approximation is qualitatively correct in a certain regime (weak coupling). Attempts to push out of this regime through higher approximations have led to great difficulties. Having learned the predictions of the theory, therefore, we should be prepared to have to seek a new method in order to understand intermediate and strong coupling.

The basic Hamiltonian H of the problem is

$$H = -\frac{\hbar^2}{2m} \sum_{i=1}^{N} \nabla_i^2 + \frac{1}{2} \sum_{i \neq j} v(\mathbf{x}_i - \mathbf{x}_j) \tag{A.1}$$

wherein v is the two–body potential (no one has yet considered systems with more–body forces), $\mathbf{x}_i$ is the coordinate of the ith particle, and N is the number of bosons. The potential v is necessarily symmetric (i.e., $v(\mathbf{x}) = v(-\mathbf{x})$), but it is not necessary that it be spherically symmetric, although this assumption is generally made. The scene of action is a box of volume $V = L \times L \times L$ and $\rho = N/V$ is the density of particles. We are interested in the bulk limit: $N \to \infty$ with $\rho =$ constant.

The first question that arises is that of boundary conditions, and this is intimately connected with the question of extensivity (or saturation in the terminology of nuclear physics). The ground state energy, E_0, is said to be extensive if it is of the form $E_0 = N \times$ function of ρ. Similarly, extensivity is defined for the free energy, F, at some non–zero temperature. The system is said to be extensive if both E_0 and F are. Unfortunately, necessary and sufficient conditions on v that the system be extensive are far from known, but one can prove extensivity for a wide class of potentials. For some of these, in turn, one can further show that E_0 and F are independent of boundary conditions (to leading order in N), provided they are fixed homogeneous conditions. We shall assume that our v is of this kind and shall use periodic boundary conditions which, while they may be somewhat unphysical, are mathematically most convenient. They state that if $\psi(\mathbf{x}_1, \ldots, \mathbf{x}_N)$ be an eigenfunction of H and if $\mathbf{x}_j$ (for all j) be on the wall of the box (so that $\mathbf{x}_j = (x, y, 0)$ for example), then the value of ψ on the opposite wall ($\mathbf{x}_j = (x, y, L)$) must be the same as on the first wall. This is to be true for all values of $\mathbf{x}_k$ ($k \neq j$), and a similar condition is imposed on the normal derivative of ψ.

Periodic conditions can *always* be imposed, but they become *useful* only if H can be periodically extended to all of space. This in turn requires that $v(\mathbf{x}_1 - \mathbf{x}_2)$ be periodic in $\mathbf{x}_1$ and $\mathbf{x}_2$, which means, in effect, that v must depend upon L. Thus, the one–dimensional potential $v(x_2 - x_1) = \exp(-\gamma|x_2 - x_1|)$ is not periodic for x_1 and x_2 in $(0, L)$, but it can be replaced by $\exp(-\gamma|x_2 - x_1|) + \exp(-\gamma L + \gamma|x_2 - x_1|)$ which is periodic in this region. It can then be periodically extended to all of space and one can show that the addition to v will not affect E_0 or F in the bulk limit.

The virtue of having H periodic is that one can trivially show that $\psi(\mathbf{x} + \mathbf{a})$ is an eigenfunction with the same energy as ψ (by this we mean that any constant vector $\mathbf{a}$ is added to all coordinates). Since all irreducible representations of the translation group are one–dimensional, every eigenfunction *can be* chosen to have a constant total momentum, i.e.,

$$\mathbf{P}\psi = -i\hbar\left(\sum_{i=1}^{N} \nabla_i\right)\psi = \mathbf{p}\psi \tag{A.2}$$

where $\mathbf{p}$ is a real vector. Another advantage is that for $\mathbf{x}_1$ and $\mathbf{x}_2$ in V, the potential can be written as

$$v(\mathbf{x}_2 - \mathbf{x}_1) = V^{-1} \sum_{\mathbf{k}} \nu(\mathbf{k})e^{i\mathbf{k}\cdot(\mathbf{x}_2 - \mathbf{x}_1)} \tag{A.3}$$

where the $\mathbf{k}$ are vectors of the form $(2\pi/L)(n_1, n_2, n_3)$, (the n_j are integers). The point here is that any potential can always be written as a double Fourier series, but the series is diagonal as in (A.3) only when v is periodic. The Fourier transform is given by

$$\nu(\mathbf{k}) = \int_V v(\mathbf{x})e^{-i\mathbf{k}\cdot\mathbf{x}}d\mathbf{x}. \tag{A.4}$$

Now the essence of the Bogoliubov method is that we go into momentum space. To do this we define

$$\varphi(\mathbf{k}_1, \dots, \mathbf{k}_N) = \int_V d\mathbf{x}_N \cdots \int_V d\mathbf{x}_1 \phi(\mathbf{x}_1, \dots, \mathbf{x}_N) \exp\left(i \sum_{i=1}^{N} \mathbf{k}_i \cdot \mathbf{x}_i\right) \tag{A.5a}$$

$$\psi(\mathbf{x}_1, \dots, \mathbf{x}_N) = V^{-N} \sum_{\mathbf{k}_1} \cdots \sum_{\mathbf{k}_N} \varphi(\mathbf{k}_1, \dots, \mathbf{k}_N) \exp\left(-i \sum_{i=1}^{N} \mathbf{k}_i \cdot \mathbf{x}_i\right) \tag{A.5b}$$

and the equation $H\psi = E\psi$ becomes

$$\left[\sum_{i=1}^{N} \epsilon(\mathbf{k}_i)\right] \varphi(\mathbf{k}_1, \dots, \mathbf{k}_N) + (1/2V) \sum_{\mathbf{p}} \sum_{i\neq j} \varphi(\mathbf{k}_1, \dots, \mathbf{k}_i + \mathbf{p},$$

$$\cdots \mathbf{k}_j - \mathbf{p}, \dots, \mathbf{k}_N)\nu(\mathbf{p}) = E\varphi(\mathbf{k}_1, \dots, \mathbf{k}_N) \tag{A.6}$$

where $\epsilon(\mathbf{k}) = (\hbar^2/2m)\mathbf{k}^2$.

Since we are interested in bosons we require solutions, φ, to (A.6) which are symmetric in $\mathbf{k}_1, \dots, \mathbf{k}_N$. Since φ is symmetric, the sums on i and $i \neq j$ in (A.6) are to some extent repetitive and it is convenient to introduce a device to handle automatically the factors of N, $\binom{N}{2}$, etc. that will continually appear on the left hand side of (A.6).

The device used is to introduce boson creation and annihilation operators. But we wish to emphasize strongly that they are not essential to the theory. They are merely a convenient bookkeeping device. For every $\mathbf{k}$ one introduces a creation operator $a_{\mathbf{k}}^\dagger$ and its Hermitian conjugate (the annihilation operator) $a_{\mathbf{k}}$ which satisfy the following commutation rules:

$$[a_{\mathbf{k}}, a_{\mathbf{k}'}] = 0 = [a_{\mathbf{k}}^\dagger, a_{\mathbf{k}'}^\dagger]$$

$$[a_{\mathbf{k}}, a_{\mathbf{k}'}^\dagger] = \delta_{\mathbf{k},\mathbf{k}'} \tag{A.7}$$

The δ is a Kronecker delta since the $\mathbf{k}$'s are discrete. A vacuum state $|0\rangle$ in a Hilbert space is also introduced such that

$$a_{\mathbf{k}}|0\rangle = 0 \qquad \text{(for all } \mathbf{k}).\qquad (A.8)$$

One can show that the state $|0\rangle$ is essentially unique and that any state in the Hilbert space is a sum of products of various $a^\dagger$'s acting on the vacuum. Next one defines an isomorphism between the functions φ and states Ψ in the Hilbert space as follows:

$$\Psi = \sum_{\mathbf{k}_N} \cdots \sum_{\mathbf{k}_1} \varphi(\mathbf{k}_1, \ldots, \mathbf{k}_N) a_{\mathbf{k}_1}^\dagger \cdots a_{\mathbf{k}_N}^\dagger |0\rangle. \qquad (A.9)$$

Using (A.7) it is easy to see that knowing Ψ one can find φ, so that the correspondence is indeed one to one. It is then slightly tedious, but simple, to show that if φ satisfied (A.6) then Ψ satisfies

$$H\Psi \equiv \left\{ \sum_{\mathbf{k}} \epsilon(\mathbf{k}) a_{\mathbf{k}}^+ a_{\mathbf{k}} + (1/2V) \sum_{\mathbf{k},\mathbf{q},\mathbf{p}} a_{\mathbf{k}+\mathbf{p}}^+ a_{\mathbf{q}-\mathbf{p}}^+ a_{\mathbf{k}} a_{\mathbf{q}}\, \nu(\mathbf{p}) \right\} \Psi = E\Psi. \qquad (A.10)$$

Equation (A.10) is the starting point for Bogoliubov's approximation.

Before proceeding it is well to keep in mind certain properties of the $a_{\mathbf{k}}$'s. We define the total number operator

$$\eta = \sum_{\mathbf{k}} a_{\mathbf{k}}^\dagger a_{\mathbf{k}} \qquad (A.11)$$

and the total momentum operator

$$\mathbf{P} = \hbar \sum_{\mathbf{k}} \mathbf{k} a_{\mathbf{k}}^\dagger a_{\mathbf{k}}. \qquad (A.12)$$

These commute with H and when acting on Ψ yield N and $\mathbf{p}$ respectively.

Now, consider the state $a_{\mathbf{k}}\Psi$. This state has particle number $N-1$ and momentum $\mathbf{p} - \hbar\mathbf{k}$. We can go back through (A.9) and (A.5) and ask what function this corresponds to in configuration space. The result is

$$a_{\mathbf{k}}\Psi \Longleftrightarrow N \int_V \Psi(\mathbf{x}_1, \ldots, \mathbf{x}_N) e^{-i\mathbf{k}\cdot\mathbf{x}_N} d^3\mathbf{x}_N. \qquad (A.13)$$

Likewise, the state $a_{\mathbf{k}}^\dagger \Psi$ has $N+1$ particles and momentum $\mathbf{p} + \hbar\mathbf{k}$. It corresponds to

$$a_{\mathbf{k}}^\dagger \Psi \Longleftrightarrow \frac{1}{(N+1)V} \sum e^{i\mathbf{k}\cdot\mathbf{x}_{N+1}} \Psi(\mathbf{x}_1, \ldots, \mathbf{x}_N) \qquad (A.14)$$

where the sum is on $N + 1$ similar terms.

It is also convenient to define density operators which are generalizations of (A.11), namely

$$\rho(\mathbf{k}) = \sum_{\mathbf{q}} a^\dagger_{\mathbf{k+q}} a_{\mathbf{q}} ; \qquad (A.15)$$

it conserves particle number and increases the momentum by an amount $\hbar\mathbf{k}$. Its effect in configuration space is given by using (A.13) and (A.14) and is

$$\rho(\mathbf{k})\Psi \Longleftrightarrow \left(\sum_{j=1}^{N} e^{i\mathbf{k}\cdot\mathbf{x}_j} \right) \Psi(\mathbf{x}_1, \ldots, \mathbf{x}_N) . \qquad (A.16)$$

We shall have occasion to use the relations (A.13) to (A.16) later.

Returning to (A.10), the reason for going into momentum space is the following: If there were no interaction then $\Psi_0 = (N!)^{1/2}(a^\dagger_0)^N |0\rangle$ would be the normalized ground state with energy zero. The interaction v has the property that it converts a pair of particles with momenta $\mathbf{p}$ and $\mathbf{q}$ into a pair with momenta $\mathbf{p} + \mathbf{k}$ and $\mathbf{q} - \mathbf{k}$. The matrix element is $(1/2V)v(\mathbf{k})$. Starting with all N particles having momentum zero (so-called condensed state), we would first get $(N - 2)$ with momentum zero, together with one pair having momenta $\mathbf{k}$ and $-\mathbf{k}$. When the potential is applied again we could get two possibilities: one would be two pairs $\mathbf{k}, -\mathbf{k}$ and $\mathbf{q}, -\mathbf{q}$; the other would be a genuine triplet $\mathbf{k}, \mathbf{q}, \mathbf{r}$, such that $\mathbf{k} + \mathbf{q} + \mathbf{r} = 0$. But the probability of the former relative to the latter would be $(N - 2)(N - 3)/4$ because there are $(N - 1)$ particles with zero momentum and only 2 with non–zero momentum. Applying v over and over again we will ultimately get a finite fraction of triplets, quartets, etc. as well as pairs, but hopefully if the interaction is weak enough we need consider explicitly only pairs in the ground state wave function. Stating this more precisely, we suppose that to a good approximation Ψ_0 is a sum of terms each of which contains several factors like $a^\dagger_{\mathbf{k}} a^\dagger_{-\mathbf{k}}$, as well as $a^\dagger_0$, acting on $|0\rangle$.

Another way of motivating this *ansatz* is to note that in the non–interacting ground state $n_{\mathbf{k}} = \langle a^\dagger_{\mathbf{k}} a_{\mathbf{k}} \rangle = N\delta_{\mathbf{k},0}$, that is to say all the particles are condensed. With a weak interaction we suppose that n_0/N is still a number of order unity and that the remaining fractions are largely grouped into pairs, for it is only pairs that can give rise to triplets. The idea that n_0/N is of order unity is called the condensation hypothesis. It need not be true for sufficiently strong interaction and we remark that Girardeau [Gi3] has generalized this concept somewhat.

If Ψ as given by (A.7) has only pairs (more precisely, for every $a^\dagger_{\mathbf{k}}$ with $\mathbf{k} \neq 0$, there is an $a^\dagger_{-\mathbf{k}}$) then only certain parts of H result in pair functions when applied to Ψ. There are three possibilities as far as the interaction is concerned: The first is when all indices are zero, giving $(a_0^\dagger)^2(a_0)^2$; the second is when two indices are zero, giving $(a_0^\dagger)^2 a_k a_{-k}$ or $a_k^\dagger a_{-k}^\dagger (a_0)^2$ or $a_k^\dagger a_0^\dagger a_{-k} a_0$; the third is when no indices are zero, giving $a_{-q}^\dagger a_q^\dagger a_k a_{-k}$ or $a_k^\dagger a_q^\dagger a_k a_q$. Collecting together all such parts of H we derive the pair Hamiltonian,

$$
\begin{aligned}
H_{\text{pair}} = {} & \frac{1}{2}(n-1)\rho v(0) + {\sum_k}' \big[\epsilon(\mathbf{k}) + (1/V)N_0 \nu(\mathbf{k}) \big] N_{\mathbf{k}} \\
& + (1/2V) {\sum_k}' \nu(\mathbf{k}) \big[\alpha_{\mathbf{k}}^\dagger \alpha_0 + \alpha_0^\dagger \alpha_{\mathbf{k}} \big] \\
& + (1/2V) {\sum_{k,q}}' \nu(\mathbf{k}) \big[\alpha_{\mathbf{q}}^\dagger \alpha_{\mathbf{q-k}} + N_{\mathbf{q-k}} N_{\mathbf{q}} \big]
\end{aligned}
\tag{A.17}
$$

where $\alpha_{\mathbf{k}} = a_{\mathbf{k}} a_{-\mathbf{k}}$, $N_{\mathbf{k}} = a_{\mathbf{k}}^\dagger a_{\mathbf{k}}$ and the prime on the summation means we delete the terms $\mathbf{k} = \mathbf{0}$ and/or $\mathbf{q} = \mathbf{0}$ as well as the term $\mathbf{q} = \mathbf{k}$ in the double summation (note that N_0 and α_0 are operators). In deriving (A.17) we used the fact that $\eta\Psi = N\Psi$.

It is important to note that H_{pair} has a *double* significance. On the one hand if we can diagonalize it we should have a good approximation to the ground state and low lying states of the system for the reasons mentioned above. On the other hand we have seen that if we take the expectation value of the total H with respect to *any* state having *only* pairs then H_{pair} is the only part of H that contributes to the final result. Hence, from the variational theorem, the exact ground state energy, $E_{0,\text{pair}}$, of H_{pair} is a true upper bound to the ground state energy, E_0, of H. Moreover, any variational upper bound to $E_{0,\text{pair}}$ is thus an upper bound to E_0. It turns out that $E_{0,\text{pair}}$ can indeed be found if one is prepared to solve a finite set of non–linear integral equations. This can be done in certain limiting cases and has been exploited by Girardeau and Arnowitt, [GA, Gi1, Gi4, Gi5].

Basically, what permits us to find the ground state energy of (A.17) in the bulk limit, as well a the free energy for non-zero temperature, is the following observation. What we have in (A.17) are bilinear forms in operators whose expectation values we believe to be extensive. Consider, for instance

$$
{\sum_{q \neq k}}' N_{\mathbf{q}} \left({\sum_k}' \nu(\mathbf{k}) N_{\mathbf{q-k}} \right).
$$

The operator in parenthesis (call it $F_{\mathbf{q}}$) we believe has an expectation value of order N (call it $N f_{\mathbf{q}}$). The root-mean-square fluctuation of $F_{\mathbf{q}}$ in the ground state ought to be of order $\sqrt{N}$, and if so, replacing F by Nf in (A.17) should make no difference to the energy to order N. It is possible to make this argument more precise [Gi5, Lu, W, BZT] by formally expanding the operators about their mean values in a power series in N^{-1}. The difficulty is that no one has shown that these series converge and they might well not. Nevertheless, the energy (or free energy) obtained under the assumption can be shown to be a genuine variational energy and so is a true upper bound.

Having replaced F_q by $N f_q$ we do the same thing with the term corresponding to α_{q-k}, N_0, α_0 and $\alpha_0^\dagger$. We are then left with a Hamiltonian involving only quadratic expressions in the a's and $a^\dagger$'s. This can be diagonalized in the standard way. The ground state wave function will then depend upon the c-numbers f_q, etc. as parameters. We then adjust these parameters so that the expectation value of F_q is indeed $N f_q$, etc. This leads to integral equations and it is clear that what we have really done is a self-consistent field calculation that we hope is rigorously correct in the bulk limit.

There is, however, still one difficulty which we have glossed over. If we replace α_0 by a c-number then the expression $\alpha_k^\dagger \alpha_0$ will no longer conserve particles because $\alpha_k^\dagger$ always creates two particles. Another way of saying this is that the expectation value of α_0 in the true ground state is, strictly speaking, zero. There are two ways around this difficulty. The first is the method used by Bogoliubov, namely to introduce a chemical potential. We write

$$H'_{\text{pair}} = H_{\text{pair}} - \mu\eta \,,$$
$$(\text{or } H' = H - \mu\eta) \,. \tag{A.18}$$

We then diagonalize H'_{pair} by replacing α_0, etc. by c-numbers and in addition to the above consistency conditions we choose μ by requiring that $\langle N \rangle = N$ in the ground state. To calculate the free energy we must use a grand-canonical ensemble. This method can be justified to the same extent as in the above discussion. In the second method [Gi5, KB] we redefine operators so that particles are conserved even after the c-number substitution. Following Kromminga and Bolsterli we introduce operators $b_k = a_0^\dagger (N_0 + 1)^{-1/2} a_k$ and their conjugates. It is then easy to show that $b_k^\dagger b_q = a_k^\dagger a_q$ (all $\mathbf{k}, \mathbf{q}$) and that $b_k b_q^\dagger = a_k a_q^\dagger$ (if $\mathbf{k} \neq \mathbf{0}$ and $\mathbf{q} \neq \mathbf{0}$ and if this operator does not act on a state with $N_0 = 0$). An annoying operator such as $\alpha_k^\dagger \alpha_0$ is then equal to $b_k^\dagger b_{-k}^\dagger (N_0(N_0 - 1))^{1/2}$ on any state. Also $\alpha_k \alpha_0^\dagger = (N_0(N_0 - 1))^{1/2} b_k b_{-k}$ and $\alpha_q^\dagger \alpha_k = b_q^\dagger b_{-q}^\dagger b_k b_{-k}$. In addition one easily proves that the b's satisfy the same commutation relations (A.7) as the a's (provided $\mathbf{k} \neq \mathbf{0}$) and also the b's conserve particle number (i.e., they commute with N). Thus either H_{pair} or H may be rewritten in terms of the b's which behave just like the a's except that now any c-number substitution will automatically preserve particle number.

All the sums explicitly exclude $\mathbf{k} = \mathbf{0}$, so the only way in which $a_{\mathbf{0}}$ and $a_{\mathbf{0}}^{\dagger}$ appear in the new Hamiltonian is through $N_{\mathbf{0}}$. But $N_{\mathbf{0}}$ may be eliminated in favor of b's by the relation

$$N_{\mathbf{0}} = N - {\sum}' a_{\mathbf{k}}^{\dagger} a_{\mathbf{k}} = N - {\sum}' b_{\mathbf{k}}^{\dagger} b_{\mathbf{k}} \,.$$

Now the particle number appears explicitly in H, which it did not before (although it did appear in H_{pair}), and a term such as $N_{\mathbf{0}} N_{\mathbf{k}}$ becomes now a quartic form in the b's.

This is a neat trick to overcome the problem of particle conservation that plagued previous authors (such as Bogoliubov). It obviates the need for lengthy (and unrigorous) arguments that method (A.18) gives the correct answer in the bulk limit. Needless to say, however, in every calculation anyone has ever done, the more cumbersome method 2 does indeed give the same result as method 1 and we shall therefore use the latter.

Bogoliubov did not actually use H_{pair}. He made a further simplification which consisted in deleting the last sum in (A.17) on the grounds that these terms are quadratic pair operators and so may be expected to be small in comparison with the first two sums. This omission unfortunately destroys the bounding property of the Hamiltonian, but it does turn out that for sufficiently weak interaction Bogoliubov's ground state energy is indeed an upper bound to E_0 as was to be expected. To that extent his simplification is justified.

We begin by replacing $N_{\mathbf{0}}$, $a_{\mathbf{0}}$ and $a_{\mathbf{0}}^{\dagger}$ by a (common) c-number, $N_{\mathbf{0}}$.[1] For weak interaction we expect this number to be close to N, and indeed it turns out to be so. The correction, $N - N_{\mathbf{0}}$, (the so-called ground state depletion) gives a higher order contribution to E_0 — a correction which is of the same order as that caused by the neglected quartic terms. Since the qualitative results do not depend upon the ground state depletion effect, we shall simply take $N_{\mathbf{0}}$ to be N. We thus have Bogoliubov's Hamiltonian (replacing $N - 1$ by N in the bulk limit):

$$H_B = \frac{1}{2} N \rho \nu(\mathbf{0}) + {\sum_{\mathbf{k}}}' f(\mathbf{k}) a_{\mathbf{k}}^{\dagger} a_{\mathbf{k}} + \frac{1}{2} g(\mathbf{k}) \left(a_{\mathbf{k}} a_{-\mathbf{k}} + a_{-\mathbf{k}}^{\dagger} a_{\mathbf{k}}^{\dagger} \right), \qquad \text{(A.19)}$$

where

$$f(\mathbf{k}) = \epsilon(\mathbf{k}) + \rho \nu(\mathbf{k})$$
$$g(\mathbf{k}) = \rho \nu(\mathbf{k}) \,. \qquad\qquad\qquad \text{(A.20)}$$

---

[1]Subsequently, Ginibre [Gin] proved that one can replace $a_{\mathbf{0}}$ and $a_{\mathbf{0}}^{\dagger}$ everywhere in H by the c-number $\sqrt{N_{\mathbf{0}}}$ without making an error in the ground state energy/particle in the thermodynamic limit. A simpler and more general version of this result can be found in Appendix D.

This Hamiltonian is a quadratic form in the a's and may be diagonalized in the usual way. The transformation that accomplishes this is $\exp{(iS)}$, where

$$iS = \frac{1}{2}\sideset{}{'}\sum_{\mathbf{k}}\left(a_{\mathbf{k}}^{\dagger}a_{-\mathbf{k}}^{\dagger} - a_{\mathbf{k}}a_{-\mathbf{k}}\right)\psi(\mathbf{k}),\qquad(A.21)$$

so that

$$a_{\mathbf{k}} \rightarrow b_{\mathbf{k}} = e^{iS}a_{\mathbf{k}}e^{-iS} = a_{\mathbf{k}}\cosh\psi(\mathbf{k}) - a_{\mathbf{k}}^{\dagger}\sinh\psi(\mathbf{k}).\qquad(A.22)$$

If we now choose

$$\tanh 2\psi(\mathbf{k}) = g(\mathbf{k})/f(\mathbf{k}),\qquad(A.23)$$

then

$$H_B \rightarrow H_B' = \frac{1}{2}N\rho\nu(0) - \frac{1}{2}\sideset{}{'}\sum_{\mathbf{k}}f(\mathbf{k}) - \left(f(\mathbf{k})^2 - g(\mathbf{k})^2\right)^{1/2}$$
$$+ \sideset{}{'}\sum_{\mathbf{k}}b_{\mathbf{k}}^{\dagger}b_{\mathbf{k}}\epsilon'(\mathbf{k}),\qquad(A.24a)$$

where

$$\epsilon'(\mathbf{k}) = \left[\epsilon(\mathbf{k})^2 + 2\epsilon(\mathbf{k})\,\rho\nu(\mathbf{k})\right]^{1/2}.\qquad(A.24b)$$

Notice that this transformation is impossible unless $|g| < |f|$. It is also necessary that $f(\mathbf{k}) > 0$ for all $\mathbf{k}$. Unless these two conditions are fulfilled the Hamiltonian has no ground state and it would then be unphysical. This means that $\nu(\mathbf{k})$ cannot be too negative (attractive), but it can be as repulsive as we please.

Now let us consider the implications of (A.24). Since the b's are bosons, the ground state wave function is given by Ψ_0', the vacuum of the b's, i.e.,

$$b_{\mathbf{k}}\Psi_0' = 0 \qquad \text{all } \mathbf{k} \neq \mathbf{0}.\qquad(A.25)$$

The ground state energy is then simply

$$E_0 = \tfrac{1}{2}N\rho\nu(0) - N\left(4\pi^2\rho\right)^{-1}\int_0^{\infty}k^2dk\{\epsilon(\mathbf{k}) + \rho\nu(\mathbf{k}) - \left(\epsilon(\mathbf{k})^2 + 2\epsilon(\mathbf{k})\rho\nu(\mathbf{k})\right)^{1/2}\},\qquad(A.26)$$

where we have gone to the bulk limit by replacing

$$\sum_{\mathbf{k}} \quad \text{by} \quad (L/2\pi)^3\int d^3k,$$

and have further assumed that the problem is spherically symmetric. At first sight it appears that the second term in (A.26) is order ρ^{-1}, thereby violating our intuition that it should be small for low density. This is not so because as $\rho \rightarrow 0$

the integrand itself vanishes. To see what happens let us assume that $\nu(\mathbf{k})$ goes to zero for large $\mathbf{k}$ faster than k^{-1}. Let us rewrite the integral in (A.26) as follows: $I = I_1 + I_2$ where

$$I_1 = \int_0^\infty k^2 dk \{\alpha^2 k^2 + \rho\nu(k) - (\alpha^4 k^4 + 2\alpha^2 k^2 \rho\nu(k))^{1/2} - \nu(k)^2/2\alpha^2 k^2\}$$

and

$$I_2 = \rho^2/2\alpha^2 \int_0^\infty \nu(k)^2 dk \tag{A.27}$$

where $\alpha^2 = \hbar^2/2m$. The integral I_2 certainly converges. In I_1 the integrand goes to zero faster than k^{-3} and is absolutely convergent. We see that as $\rho \to 0$ only very small k will play a role. Let us assume that $\nu(k)$ is smooth for small k and that $\nu(0) \neq 0$. We may then replace $\nu(k)$ by $\nu(0)$ everywhere and the integral is then elementary.

Let us define

$$a_0 = (8\pi\alpha^2)^{-1}\nu(0),$$

and

$$a_1 = -(2\pi^2)^{-1}(8\pi\alpha^2)^{-2} \int d^3k\, \nu(k)/k^2. \tag{A.28}$$

The result for E_0 is then

$$E_0 = 4\pi N\rho\,(\hbar^2/2m)(a_0 + a_1) + 4\pi N\rho\,(\hbar^2/2m)\,a_0(128/15\sqrt{\pi})(\rho a_0^3)^{1/2}. \tag{A.29}$$

What is the significance of this result? If we had neglected the integral in (A.26) we would have gotten $E_0 = \frac{1}{2}N\rho\nu(0) = 4\pi N\rho\hbar^2(2m)^{-1}a_0$. We may call this the zeroth order Bogoliubov approximation. But notice that it really does not depend upon Planck's constant and the mass – a conclusion that is certainly meaningless because if the mass were infinite E_0 would be equal to the minimum potential energy which is not necessarily $N\rho\nu(0)$. The integral term comes to the rescue, however. We had naively expected it to contribute a higher power of the density than ρ, but it in fact contributed a term of the same order as the zeroth approximation – namely ρa_1. This term now truly depends on α^2. If we look a little closer we notice that $a_0 + a_1$ are just the first two terms in the Born series for the scattering length. That is to say, if we consider the zero–energy scattering equation

$$\left[-\alpha^2\left(\nabla_1^2 + \nabla_2^2\right) + v(\mathbf{x}_1 - \mathbf{x}_2)\right]\psi(\mathbf{x}_1 - \mathbf{x}_2) = 0, \tag{A.30}$$

the asymptotic behavior of ψ is $\psi(\mathbf{x}_1 - \mathbf{x}_2) \sim 1 - a/|\mathbf{x}_1 - \mathbf{x}_2|$. The quantity a is defined to be the scattering length. (From (A.30), it is also given by $a = (8\pi\alpha^2)^{-1} \int v(\mathbf{x})\psi(\mathbf{x})d^3\mathbf{x}$.) We therefore suspect, despite what we had originally thought, that Bogoliubov's method is really an expansion in the density *and* in the potential v. It is not truly a low density expansion unless the potential is very weak. This idea has been confirmed by doing perturbation theory on the parts of H not included in (A.19), using our wave function previously found as the starting point. It is indeed true that higher order corrections give contributions proportional to ρ. They can be recognized as constituting the full Born series for the scattering length. Likewise, the second term in (A.19) (which came from I_1) looks like it is the beginning of a similar series. Hence we are tempted to write

$$E_0 = (\hbar^2/2m)4\pi N\rho\, a\left\{1 + (128/15\sqrt{\pi})\,(\rho a^3)^{1/2} + \cdots\right\} \qquad (A.31)$$

and presumably we now have the beginning of a genuine series in the density *alone*.

Nevertheless, it is important to note that what started out to be a very reasonable hypothesis – the pair approximation – is invalid as a density expansion. Even the full pair Hamiltonian (A.17) will not give (A.31) [Gi1]. The trouble was that we had thought we were making some sort of cluster expansion as one does in classical statistical mechanics. This may be a reasonable thing to do, but it is essential that we treat the two–body interaction fully and completely, and this cannot be done very easily by perturbation theory. We shall discuss this matter more fully later but for the present let us make some attempt to justify (A.31).

The expression (A.31) for E_0 has not yet been proved to be correct. But it is very reasonable. For one thing the first term is simply the number of pairs of particles, $\frac{1}{2}N(N-1)$, times the ground state energy of two particles in a large box, $\alpha^2 8\pi a/V$.[2] The second term is harder to understand. There is no analogue of it for two particles and it is clearly some sort of quantum-mechanical correlation effect. Nevertheless, if the true second order term in the density is of the form $\rho^{3/2}$ given by (A.31) it must, for dimensional reasons, be proportional to a $(\text{length})^{5/2}$. But the only relevant length at low energy is the scattering length.

Having considered the ground state energy let us return to the second term in (A.24). Apart from the fact that ϵ' is different from ϵ, this spectrum of H_B is the same as for the original Hamiltonian without interaction. Any number, n, of bosons (or phonons) of any momentum $\mathbf{k}$ can be excited independently of each other with energy $n\epsilon'(\mathbf{k})$. The important difference is that for low momentum $\epsilon(\mathbf{k})$ is proportional to $\mathbf{k}^2$, whereas $\epsilon'(\mathbf{k}) = [2\alpha^2\rho v(0)]^{1/2}|\mathbf{k}|$. This new spectrum is

---

[2]This is discussed in Chapter 2 in connection with Eq. (2.13).

definitely phonon-like (without an energy gap) and we expect that it is associated with sound propagation. If so, the velocity of sound would be

$$v_s = \lim_{k \to 0} (1/\hbar k)\epsilon'(k) = (\rho \nu(0)/m)^{1/2} = (\hbar/m)(4\pi \rho a_0)^{1/2}. \tag{A.32}$$

We can check this result by using the fundamental definition of the velocity of sound in terms of the compressibility,

$$v_s = \left[m^{-1}(\partial/\partial \rho) \, \rho^2 \partial/\partial \rho \big(E_0/N\big) \right]^{1/2}.$$

Using (A.29) this gives

$$v_s = \hbar/m \left[4\pi\rho(a_0 + a_1) + 64\rho \, a_0 \big(\pi \, \rho \, a_0^3\big)^{1/2} \right]^{1/2}. \tag{A.33}$$

There is agreement between (A.33) and (A.32) and we are led to surmise as before that the correct expression to the first two orders in the density is

$$v_s = \hbar/m \left[4\pi \, \rho \, a \big(1 + (16/\pi)(\pi \, \rho \, a^3)\big)^{1/2} \right]^{1/2}. \tag{A.34}$$

It is interesting to note that the expression (A.33), obtained from E_0, is more accurate than (A.32) obtained from the phonon spectrum. This is curious since we would have expected that whatever accuracy was inherent in H_B, it should be the same for $\epsilon'(\mathbf{k})$ as for E_0.

Another important feature of the boson type spectrum that we have obtained is that while there may be something qualitatively correct about it, it is much too simple to be taken literally. It is hardly to be expected that the spectroscopy of the true spectrum will fall into a pattern associated with independent normal modes. It must be more complicated and indeed higher order perturbation theory indicates that the phonons interact with one another. There are two ways to describe this state of affairs. The usual way is to say that the interaction causes the phonons to decay with a finite lifetime. The second way is to say that the unitary transformation leading to the b_k's was only a partial diagonalization of the full Hamiltonian which must always have real eigenvalues since it is Hermitian. In other words the true energy spectrum and eigenfunctions are simply more complicated than we have so far envisaged. Still, the independent phonon idea may be justified provided we do not excite too many of them – this would be true at low temperature.

We might, however, anticipate a difficulty of another sort. The Bogoliubov approximation is essentially perturbation theory, albeit of a sophisticated sort, because it assumes that the system does not change drastically when we switch on the interaction. Were it otherwise H_B could not be justified. It is generally held that if we imagine $v(\mathbf{x})$ to be proportional to a coupling constant λ, then after passing to the bulk limit there will be an analytic singularity in λ (in $E_0(\lambda)$ for example) at $\lambda = 0$. A Bose gas with an interaction, however weak, may be qualitatively different from the non-interacting gas. If this is so then we might expect new types of normal modes (or phonons) that have no counterpart for free bosons. Indeed, when we examine a one-dimensional model later we shall see that to the extent that the spectrum is phonon like, it can best be described by two separate $\epsilon(\mathbf{k})$ curves, not one.

Where might this second phonon spectrum come from? In the Bogoliubov ground state most of the particles are still condensed at $\mathbf{k} = \mathbf{0}$ and there is a small (but non-zero) background of particles with $\mathbf{k} \neq \mathbf{0}$. The Bogoliubov phonons are qualitatively the same as the free "phonons"; particles are excited out of $\mathbf{k} = \mathbf{0}$ sea. The renormalized energy ϵ' comes about because the $\mathbf{k} = \mathbf{0}$ background now has to readjust itself. It is quite possible that there is another type of excitation which would be associated directly with excitation of the $\mathbf{k} \neq \mathbf{0}$ background. Similar suggestions have been made before (see [Ma]).

A physical reason for the possibility of another spectrum is the critical velocity failure of current theory. Some time ago Landau gave an argument to account for superfluidity based on the phonon hypothesis. Suppose one has a mass, M, of fluid moving with velocity v and momentum $P = Mv$. Let us suppose that in the rest system of the fluid it is in its ground state and hence has momentum equal to zero. In the laboratory system the fluid will have energy $E_1 = E_0 + P^2/2M$. If the fluid interacts with the walls of the channel in which it is moving it must *lose* energy and momentum, so that its energy is now $E_1 - \Delta$ and its momentum is $P - \delta$. If we make a Galilean transformation with velocity $-v$, the energy will be $E_0 - \Delta + \delta v$ and the momentum will be $-\delta$. Now consider the $\epsilon'(k)$ curve. It is presumably possible to draw a straight line $\epsilon''(k) = sk$ such that ϵ'' is just tangent to, and otherwise always under, $\epsilon'(k)$ (we could not do this for free bosons unless s were to be zero). This line defines a velocity $v_c = s/\hbar$. In order to impart momentum δ to the system the energy must therefore increase by at least an amount $v_c \delta$. In the above example, however, the energy increased by an amount less than $v\delta$. The conclusion is that if the velocity of the fluid is less than v_c, the fluid will not be able to lose any momentum to the walls of the channel, and hence it will display superfluid behavior.

We would expect from the Bogoliubov solution that v_c should be of the order of magnitude of the sound velocity. Experimentally, v_c is found to be very much less than v_s, and it is also found to depend sensitively on the diameter of the channel – especially for very narrow channels. It is clear that another type of excitation, with an energy much less than that given by (A.24b), could account for the discrepancy. Indeed, we will find such a spectrum in the one–dimensional model to be analyzed later.

Finally, we wish to emphasize another important feature of Bogoliubov's analysis, namely that the excitation spectrum is intimately connected with the ground state wave function. This fact is important because it shows that calculating the ground state energy is not merely an academic exercise. Although E_0 is a rather unimportant number (it can be measured, however) it is manifestly clear that we cannot really hope to be able to predict the dynamics of this complicated system unless we are in a position to calculate the much simpler quantity, E_0, along with some of the important properties of the ground state, such as the two–particle correlation function which we shall discuss later.

We shall write down the Bogoliubov wave function $\Psi_0' = e^{-iS}\Psi_0$. S is given by (A.21), but that expression is needlessly complicated because it contains $a_\mathbf{k}a_{-\mathbf{k}}$ which vanishes when applied to $|0\rangle$. A little algebra will show that

$$\Psi_0' = \exp\left\{\sum{}' h(\mathbf{k})a_\mathbf{k}^+ a_{-\mathbf{k}}^+\right\}\Psi_0 \tag{A.35}$$

where

$$h(\mathbf{k}) = -\frac{1}{2g(\mathbf{k})}\left\{\left(f(\mathbf{k}) - (f(\mathbf{k})^2 - g(\mathbf{k})^2)^{1/2}\right)\right\}$$
$$= -\frac{1}{2} - \frac{1}{2}(\rho\nu(\mathbf{k}))^{-1}\left(\epsilon(\mathbf{k}) - \epsilon'(\mathbf{k})\right). \tag{A.36}$$

In (A.35) the state Ψ_0' is not normalized, and Ψ_0 means the free particle ground state, i.e., $(a_\mathbf{0}^\dagger)^N|0\rangle$. The difficulty with Ψ_0', of course, is that it does not have a definite particle number. One way to fix this would be to replace the $a^\dagger$'s by the Kromminga, Bolsterli $b^\dagger$'s. This would not alter E_0. A simpler procedure would be to multiply each $a_\mathbf{k}^\dagger$ by $N^{-1/2}a_\mathbf{0}$. If we then write out the first few terms in (A.35) we get

$$\Psi_0' = (a_\mathbf{0}^\dagger)^N|0\rangle + N\sum_\mathbf{k}{}' h(\mathbf{k})a_\mathbf{k}^\dagger a_{-\mathbf{k}}^\dagger(a_\mathbf{0}^\dagger)^{N-2}|0\rangle$$
$$+ N^2/2!\sum_{\mathbf{k},\mathbf{q}} h(\mathbf{k})h(\mathbf{q})a_\mathbf{k}^\dagger a_{-\mathbf{k}}^\dagger a_\mathbf{q}^\dagger a_{-\mathbf{q}}^\dagger(a_\mathbf{0}^\dagger)^{N-4}|0\rangle \tag{A.37}$$

$$+ \dots.$$

In essence, (A.37) may be taken as the statement of the Bogoliubov *ansatz*, just as much as (A.19), for the most general wave function we could construct would be

$$\Psi = (a_0^\dagger)^N|0\rangle + N\sum_k{}' h^{(2)}(\mathbf{k})a_\mathbf{k}^\dagger a_{-\mathbf{k}}^\dagger (a_0^\dagger)^{N-2}|0\rangle$$

$$+ N\sum_{\mathbf{k},\mathbf{q}}{}' h^{(3)}(k,q)a_\mathbf{k}^\dagger a_\mathbf{q}^\dagger a_{-\mathbf{k}-\mathbf{q}}^\dagger (a_0^\dagger)^{N-3}|0\rangle$$

$$+ N^2/2! \sum_{\mathbf{k},\mathbf{q},\mathbf{p}}{}' h^{(4)}(\mathbf{k},\mathbf{q},\mathbf{p})a_\mathbf{k}^\dagger a_\mathbf{q}^\dagger a_\mathbf{p}^\dagger a_{-\mathbf{k}-\mathbf{p}-\mathbf{q}}^\dagger (a_0^\dagger)^{N-4}|0\rangle \qquad \text{(A.38)}$$

$$+ \ldots$$

(Notice that there can be no linear term because of momentum conservation.) The quadratic term in (A.37) is quite general, but (A.37) has no triplet or other odd power terms. Also the quartic term is of a very special kind. The Bogoliubov *ansatz* is that

$$h^{(4)}(\mathbf{k},\mathbf{q},\mathbf{p}) = \delta_{\mathbf{p}+\mathbf{k}} h^{(2)}(\mathbf{k})h^{(2)}(\mathbf{q}), \qquad \text{(A.39)}$$

and so on for succeeding even powers. We could just as well have derived the Bogoliubov results by starting with (A.37) instead of with H_B in (A.19).

Appendix B

An Exactly Soluble Model

This appendix is taken from Section 5 of [L3], which summarizes the papers [LL, L1]. Until recently this model was just an amusing exercise, but it now appears that it is possible to produce a one-dimensional gas like this in the laboratory and the experimental results agree, so far, with the rigorous calculations on the model. In Chapter 8 we show how this one-dimensional model emerges from three-dimensional models when the trap is long compared to its width.

With various approximation schemes before us, it would certainly be advantageous to have at least one problem of the type (A.1) that can be solved exactly. We should then be able to verify whether or not the previously mentioned qualitative ideas are correct.

Such model problem is the one-dimensional Bose gas with a pair-wise repulsive δ-function potential ([LL, L1]). Using units in which $\hbar^2/2m = 1$, the Schrödinger equation is

$$\left\{ -\sum_{i=1}^{N} \partial^2/\partial x_i^2 + 2c \sum_{i<j} \delta(x_i - x_j) \right\} \psi = E\psi. \tag{B.1}$$

Hence $2c \geq 0$ is the amplitude of the δ–function. If L is the length of the line then $\rho = N/L$. It is well known that a δ–function potential is equivalent to the following boundary conditions whenever any two particles touch each other (irrespective of the value of the remaining $N - 2$ coordinates):

$$(\partial/\partial x_j - \partial/\partial x_k)\psi|_{x_j=x_k^+} - (\partial/\partial x_j - \partial/\partial x_k)\psi|_{x_j=x_k^-} = 2c\psi|_{x_j=x_k}. \tag{B.2}$$

We also note that we are seeking symmetric solutions to (B.1) and hence if we know ψ in R_1

$$R_1 : 0 \le x_1 \le \cdots \le x_N \le L, \tag{B.3}$$

we know ψ everywhere by symmetric extension. Thus our equations become

$$\left\{-\sum_{i=1}^{N} \partial^2/\partial x_i^2\right\} \psi = E\psi \ \underline{\text{inside}} \ R_1 \tag{B.4}$$

$$(\partial/\partial x_{j+1} - \partial/\partial x_j)\psi|_{x_{j+1}=x_j^+} = c\psi|_{x_{j+1}=x_j}. \tag{B.5}$$

Moreover, the original periodic boundary conditions can be interpreted as

$$\psi(0, x_2, \ldots, x_N) = \psi(x_2, \ldots, x_N, L)$$
$$\partial/\partial x \, \psi(x, x_2, \ldots, x_N)|_{x=0} = \partial/\partial x \, \psi(x_2, \ldots, x_N, x)|_{x=L}. \tag{B.6}$$

To solve these equations, consider the function

$$\varphi(x_1, \ldots, x_N) = \text{Det}|\exp(i\,k_i x_j)| \tag{B.7}$$

where $k_1, \ldots, k_N \equiv \{k\}$ are any set of N distinct numbers. Now define ψ by

$$\psi = \prod_{j>i}(\partial/\partial x_j - \partial/\partial x_i + c)\varphi. \tag{B.8}$$

It is readily verified that ψ satisfies (B.5) automatically, as well as (B.4) with

$$E = \sum_{i=1}^{N} k_i^2. \tag{B.9}$$

It is (B.6) which determines the numbers $\{k\}$ and it may be shown that this is equivalent to the N simultaneous equations

$$(-1)^{N-1}\exp(-i\,k_j L) = \exp\left[i\sum_{s=1}^{N}\theta(k_s - k_j)\right], \tag{B.10}$$

where

$$\theta(k) = -2\tan^{-1}(k/c), \qquad -\pi < \theta < \pi \qquad \text{for real } k. \tag{B.11}$$

Equation (B.10) may be rewritten as

$$\delta_j \equiv (k_{j+1} - k_j)L = \sum_{s=1}^{N}\left[\theta(k_s - k_j) - \theta(k_s - k_{j+1})\right] + 2\pi\,n_j \tag{B.12}$$

for $j = 1, \ldots, N - 1$. Equation (B.12) is $N - 1$ simultaneous equations for the δ_j; when they are found, the individual k's may be obtained from (B.10). The n_j in (B.12) are integers, and it can be shown that for any choice of the n_j such that all $n_j \geq 1$ there is a solution to (B.12) with real $\delta_j \geq 0$. Presumably these solutions are unique and are the only solutions to (B.10); at least this is true if $N = 2$. These n's are therefore the quantum numbers of the system.

If we pass now to the bulk limit, $N \to \infty$, the ground state will be obtained when all $n_j = 1$, because this choice clearly minimizes (B.9). It may be verified that all the k's must lie between $-\pi\rho$ and $\pi\rho$, which means that the spacing between the k's decreases as $N \to \infty$. If we set

$$f(k_j) = 1/\delta_j \tag{B.13}$$

then $Lf(k) =$ the number of k's between k and $k + dk$. Furthermore, we denote by K the common value of k_N and $-k_1$. Using Poisson's formula, (B.10) may be converted into an integral equation:

$$2c \int_{-K}^{K} f(p) dp \Big/ \left[c^2 + (p - k)^2 \right] = 2\pi f(k) - 1 , \tag{B.14}$$

with

$$\int_{-K}^{K} f(k) \, dk = \rho \tag{B.15}$$

being the condition that the total number of particles be N. This latter condition determines K. The ground state energy is then

$$E_0 = (N/\rho) \int_{-K}^{K} f(k) k^2 \, dk . \tag{B.16}$$

At this point it is convenient to introduce the dimensionless coupling constant

$$\gamma = c/\rho , \tag{B.17}$$

in terms of which we may write

$$E_0 = N\rho^2 e(\gamma) . \tag{B.18}$$

Equations (B.14) and (B.15) can easily be solved on a computer for all values of $\gamma \geq 0$. Graphs for $K(\gamma)$, $e(\gamma)$ and $f(k, \gamma)$ are given in Figures 1, 3 and 2 respectively of reference [LL]. The results, briefly, are these:

$$\underline{\text{small } \gamma :} \quad e(\gamma) = \gamma - (4/3\pi)\gamma^{3/2}$$
$$K(\gamma) = 2\rho\gamma^{1/2} \tag{B.19}$$
$$f(k, \gamma) = (2\pi\rho\gamma)^{-1}(4\rho^2\gamma - k^2)^{1/2}$$

$$\underline{\text{large } \gamma:} \qquad e(\gamma) = (\pi^2\gamma^2)/3)(\gamma + 2)^{-2}$$
$$K(\gamma) = \pi\rho\gamma(\gamma + 2)^{-1} \qquad\qquad \text{(B.20)}$$
$$f(k, \gamma) = (\gamma + 2)/2\pi\gamma.$$

We may inquire how Bogoliubov's theory fares for this problem. This theory yields (B.19) for $e(\gamma)$ for *all* γ, a result which is in fair agreement with the correct $e(\gamma)$ up to $\gamma = 2$ and then becomes quite useless. While the true $e(\gamma)$ is a monotonically increasing function of γ, with an asymptotic value of $\pi^2/3$, (B.19) is actually negative for $\gamma > (3\pi/4)^2$.

When γ is infinite, the wave functions, (B.8), are given by

$$\psi = \text{Det} \,| \exp(i\, k_i x_j)| \qquad\qquad \text{(B.21)}$$

where each k_i is of the form:

$$k = (2\pi/L) \times (\text{integer}), \qquad\quad (N \text{ odd})$$
$$= (2\pi/L) \times (\text{integer}) + \pi/L, \quad (N \text{ even}). \qquad \text{(B.22)}$$

For the ground state the k's run between $-\pi\rho$ and $\pi\rho$. Another way to express the ground state ψ_0 (for any N) is to recognize (B.21) as a Vandermonde determinant, whence

$$\psi_0 = \prod_{i<j} \sin(\pi|x_j - x_i|/L) \qquad\qquad \text{(B.23)}$$

for *all* values of the x_i in $(0, L)$.

Turning now to the excitation spectrum, let us first consider the infinite γ case, where the k's are given by (B.22). The spectrum will be recognized as the same as that of a one–component Fermi gas.

An elementary excitation consists in increasing one of the momenta from $q < K = \pi\rho$ to $k > K$. The energy is then

$$\epsilon(k, q) = k^2 - q^2, \qquad\qquad \text{(B.24)}$$

and the momentum of the state is

$$p(k, q) = k - q. \qquad\qquad \text{(B.25)}$$

The difficulty with this description is that it is completely different from what we had been led to expect on the basis of the pair Hamiltonian calculation. For one thing there is no unique $\epsilon(p)$ curve. For another, each excitation may take place only once, whereas a boson type excitation can be repeated as often as desired.

In order to make this spectrum appear boson-like, let us define *two* types of elementary excitations: for type I we increase k_N from K to $K + p$ (where $p > 0$). The momentum of the state is p and the energy is

$$\epsilon_1(p) = (K + p)^2 - K^2 = p^2 + 2\pi\rho\,p.$$ (B.26)

For type II we increase one of the momenta from $K - p + 2\pi/L$ (where $0 < p < K$) to $K + 2\pi/L$. Here the momentum is again p and

$$\epsilon_2(p) = 2\pi\rho\,p - p^2.$$ (B.27)

A type II excitation is defined *only* for momentum less than $\pi\rho$. With this description we have achieved our aim, but at the expense of introducing two $\epsilon(p)$ curves. Any type I excitation can be repeated as often as desired providing we agree always to take the last available k less than K and increase it by p. Similarly, with the same proviso, a type II excitation is boson-like. In addition, both I and II excitations may occur simultaneously. In fact the excitation in (B.24) may be thought of as a simultaneous type I and II excitation with momenta $p - k$ and $K - q$, respectively. It will be seen that provided we make a *finite* number of excitations by the above rule the energies and momenta will be additive to order $1/N$. Thus if we make n type I excitations with momenta $p_1, \ldots, p_n$ and m type II excitations with momenta $q_1, \ldots, q_m$, the energy as given by (B.26) and (B.27) would be

$$E = \sum_{j=1}^{n} \epsilon_1(p_j) + \sum_{j=1}^{m} \epsilon_2(q_j),$$ (B.28)

while the momentum would be

$$P = \sum_{j=1}^{n} p_j + \sum_{j=1}^{m} q_j.$$ (B.29)

Now if we examine the state we would obtain with these excitations we will find that the *true* energy of the state agrees with (B.28) to order $1/N$ while the *true* momentum is exactly given by (B.29).

There is one *caveat*, however. In achieving this boson description of the excitations we have, in reality, counted each state twice. If the rules above are carefully examined it turns out that a state with n type I excitations with momentum $p = 2\pi m/L$ is identical to the state with m type II excitations of momentum $p = 2\pi n/L$. The spectrum is therefore really much more complicated than we had imagined. If we give up the double spectrum point of view in order to avoid the double counting, then we would have to regard a type II excitation, for example, as essentially an infinite number of type I excitations with vanishing small momentum, $2\pi/L$. Not only is this unnatural, but the energy would then be given incorrectly. From (B.26) we would conclude that $\epsilon_2(p) = 2\pi\rho p$, where as the correct expression is (B.27).

When γ is not infinite the same qualitative conclusions apply. It will be appreciated that increasing one of the k's is the same as putting all but one of the n_j's in (B.12) equal to unity. For a type I excitation the singular n is n_{N-1} and this is set equal to $qL/2\pi$. For a type II excitation the singular n is one of the n_j's (where $N/2 < j < N - 1$) and this is set equal to 2. The difficulty with the finite γ case is that when the n's are changed in this way *all* the k's are shifted — not merely one of them. This shift can be computed from (B.12) and one can again obtain two $\epsilon(p)$ functions. The details are given in reference [L1]. It is found that as $\gamma \to 0$, $\epsilon_1(p) \to p^2$ (the free boson function) while $\epsilon_2(p) \to 0$ for all p. But for every γ there are always *two* $\epsilon(p)$ curves.

Bogoliubov's theory, on the other hand, predicts only *one* $\epsilon(p)$ curve:

$$\epsilon(p) = p(p^2 + 4\gamma\rho^2)^{1/2}, \quad \text{(Bogoliubov)}. \tag{B.30}$$

It turns our that for small γ, (B.30) is quite close to the true $\epsilon_1(p)$ (cf. Fig. 4 of Reference [L1]). The $\epsilon_2(p)$ curve is entirely missing from Bogoliubov's theory.

Appendix C

Definition and Properties of Scattering Length

This appendix is from the paper [LY2] and is included here because the scattering length plays an important role in these notes and because it is not easy to find a rigorous definition in the textbooks.

In this appendix we shall define and derive the scattering length and some of its properties. The reader is referred to [LLo], especially chapters 9 and 11, for many of the concepts and facts we shall use here.

We start with a potential $\frac{1}{2}v(\mathbf{x})$ that depends only on the radius, $r = |\mathbf{x}|$, with $\mathbf{x} \in \mathbb{R}^n$. For simplicity, we assume that v has finite range; this condition can easily be relaxed, but we shall not do so here, except for a remark at the end that shows how to extend the concepts to infinite range, nonnegative potentials. Thus, we assume that

$$v(r) = 0 \qquad \text{for } r > R_0. \tag{C.1}$$

We decompose v into its positive and negative parts, $v = v_+ - v_-$, with $v_+, v_- \geq 0$, and assume the following for v_- only (with $\varepsilon > 0$):

$$v_- \in \begin{cases} L^1(\mathbb{R}^1) & \text{for } n = 1 \\ L^{1+\varepsilon}(\mathbb{R}^2) & \text{for } n = 2 \\ L^{n/2}(\mathbb{R}^n) & \text{for } n \geq 3. \end{cases} \tag{C.2}$$

In fact, v can even be a finite, spherically symmetric measure, e.g., a sum of delta functions.

We also make the *important assumption* that $\frac{1}{2}v(\mathbf{x})$ has no negative energy bound states in $L^2(\mathbb{R}^n)$, which is to say we assume that for all $\phi \in H^1(\mathbb{R}^n)$ (the space of L^2 functions with L^2 derivatives)

$$\int_{\mathbb{R}^n} \left(\mu |\nabla\phi(\mathbf{x})|^2 + \frac{1}{2}v(\mathbf{x})|\phi(\mathbf{x})|^2 \right) d^n\mathbf{x} \geq 0 . \tag{C.3}$$

The scattering length is defined, of course, even when bound states are present, but it is not defined by the variational principle given below.

Theorem C.1. *Let* $R > R_0$ *and let* $B_R \subset \mathbb{R}^n$ *denote the ball* $\{\mathbf{x} : 0 < |\mathbf{x}| < R\}$ *and* S_R *the sphere* $\{\mathbf{x} : |\mathbf{x}| = R\}$. *For* $f \in H^1(B_R)$ *we set*

$$\mathcal{E}_R[\phi] = \int_{B_R} \left(\mu |\nabla\phi(\mathbf{x})|^2 + \frac{1}{2}v(\mathbf{x})|\phi(\mathbf{x})|^2 \right) d^n\mathbf{x} . \tag{C.4}$$

Then, in the subclass of functions such that $\phi(\mathbf{x}) = 1$ *for all* $\mathbf{x} \in S_R$, *there is a unique function* ϕ_0 *that minimizes* $\mathcal{E}_R[\phi]$. *This function is nonnegative and spherically symmetric, i.e,*

$$\phi_0(\mathbf{x}) = f_0(|\mathbf{x}|) \tag{C.5}$$

with a nonnegative function f_0 *on the interval* $(0, R]$, *and it satisfies the equation*

$$-\mu\Delta\phi_0(\mathbf{x}) + \frac{1}{2}v(\mathbf{x})\phi_0(\mathbf{x}) = 0 \tag{C.6}$$

in the sense of distributions on B_R, *with boundary condition* $f_0(R) = 1$.
   For $R_0 < r < R$

$$f_0(r) = f_0^{\text{asymp}}(r) \equiv \begin{cases} (r-a)/(R-a) & \text{for } n = 1 \\ \ln(r/a)/\ln(R/a) & \text{for } n = 2 \\ (1 - ar^{2-n})/(1 - aR^{2-n}) & \text{for } n \geq 3 \end{cases} \tag{C.7}$$

for some number a *called the* scattering length.
   The minimum value of $\mathcal{E}_R[\phi]$ *is*

$$E = \begin{cases} 2\mu/(R-a) & \text{for } n = 1 \\ 2\pi\mu/\ln(R/a) & \text{for } n = 2 \\ 2\pi^{n/2}\mu a/[\Gamma(n/2)(1 - aR^{2-n})] & \text{for } n \geq 3. \end{cases} \tag{C.8}$$

Remarks: 1. Given that the minimizer is spherically symmetric for every R, it is then easy to see that the R dependence is trivial. There is really one function, F_0, defined on all of the positive half axis, such that $f_0(r) = F_0(r)/F_0(R)$. That is why we did not bother to indicate the explicit dependence of f_0 on R. The reason is a simple one: If $\widetilde{R} > R$, take the minimizer $\widetilde{f_0}$ for $\widetilde{R}$ and replace its values for $r < R$ by $f_0(r)\widetilde{f_0}(R)$, where f_0 is the minimizer for the B_R problem. This substitution cannot increase $\mathcal{E}_{\widetilde{R}}$. Thus, by uniqueness, we must have that $\widetilde{f_0}(r) = f_0(r)\widetilde{f_0}(R)$ for $r \leq R$.

2. From (C.7) we then see that $f_0^{\mathrm{asymp}}(r) \geq 0$ for all $r > R_0$, which implies that $a \leq R_0$ for $n \leq 3$ and $a \leq R_0^{n-2}$ for $n > 3$.

3. According to our definition (C.7), a has the dimension of a length only when $n \leq 3$.

4. The variational principle (C.4), (C.8) allows us to discuss the connection between the scattering length and $\int v$. We recall Bogoliubov's perturbation theory [Bo, BZ], which says that to leading order in the density ρ, the energy per particle of a Bose gas is $e_0(\rho) \sim 2\pi\rho \int v$, whereas the correct formula in two-dimensions is $4\pi\mu\rho|\ln(\rho a^2)|^{-1}$. The Bogoliubov formula is an upper bound (for all ρ) since it is the expectation value of H_N in the non-interacting ground state $\Psi \equiv 1$. Thus, we must have $\frac{1}{2}\int v \geq |\ln(\rho a^2)|^{-1}$ when $\rho a^2 \ll 1$, which suggests that

$$\int_{\mathbb{R}^2} v \geq \frac{4\pi\mu}{\ln(R_0/a)}. \tag{C.9}$$

Indeed, the truth of (C.9) can be verified by using the function $\phi(\mathbf{x}) \equiv 1$ as a trial function in (C.4). Then, using (C.8), $\frac{1}{2}\int v \geq E = 2\pi\mu/\ln(R/a)$ for all $R \geq R_0$, which proves (C.9). As $a \to 0$, (C.9) becomes an equality, however, in the sense that $(\int_{\mathbb{R}^2} v)\ln(R_0/a) \to 4\pi\mu$.

In the same way, we can derive the inequality of Spruch and Rosenberg [SR] for dimension 3 or more:

$$\int_{\mathbb{R}^n} v \geq \frac{4\pi^{n/2}\mu a}{\Gamma(n/2)}. \tag{C.10}$$

(Here, we take the limit $R \to \infty$ in (C.8)).

In one-dimension we obtain (with $R = R_0$)

$$\int_{\mathbb{R}} v \geq \frac{4\mu}{R_0 - a}. \tag{C.11}$$

Proof of Theorem C.1. Given any $\phi \in H^1$ we can replace it by the square root of the spherical average of $|\phi|^2$. This preserves the boundary condition at $|\mathbf{x}| = R$, while the v term in (C.4) is unchanged. It also lowers the gradient term in (C.4) because the map $\rho \mapsto \int (\nabla \sqrt{\rho})^2$ is convex [LLo]. Indeed, there is a strict decrease unless ϕ is already spherically symmetric and nonnegative.

Thus, without loss of generality, we may consider only nonnegative, spherically symmetric functions. We may also assume that in the annular region $\mathcal{A} = \{\mathbf{x} : R_0 \leq |\mathbf{x}| \leq R\}$ there is some a such that (C.7) is true because these are the only spherically symmetric, harmonic functions in $\mathcal{A}$. If we substitute for ϕ the harmonic function in $\mathcal{A}$ that agrees with ϕ at $|\mathbf{x}| = R_0$ and $|\mathbf{x}| = 1$ we will lower $\mathcal{E}_R$ unless ϕ is already harmonic in $\mathcal{A}$. (We allow the possibility $a = 0$ for $n \leq 2$, meaning that $\phi = $ constant.)

Next, we note that $\mathcal{E}_R[\phi]$ is bounded below. If it were not bounded then (with R fixed) we could find a sequence ϕ^j such that $\mathcal{E}_R(\phi^j) \to -\infty$. However, if h is a smooth function on $\mathbb{R}_+$ with $h(r) = 1$ for $r < R + 1$ and $h(r) = 0$ for $r > 2R + 1$ then the function $\widehat{\phi^j}(\mathbf{x}) = \phi^j(\mathbf{x})$ for $|\mathbf{x}| \leq R$ and $\widehat{\phi^j}(\mathbf{x}) = h(|\mathbf{x}|)$ for $|\mathbf{x}| > R$ is a legitimate variational function for the $L^2(\mathbb{R}^n)$ problem in (C.3). It is easy to see that $\mathcal{E}_R[\widehat{\phi^j}] \leq \mathcal{E}_R[\phi^j] + (\text{const}) R^{n-2}$, and this contradicts (C.3) (recall that R is fixed).

Now we take a minimizing sequence ϕ^j for $\mathcal{E}_R$ and corresponding $\widehat{\phi^j}$ as above. By the assumptions on v_- we can see that the kinetic energy $T^j = \int |\nabla \phi^j|^2$ and $\int |\phi^j|^2$ are bounded. We can then find a subsequence of the $\widehat{\phi^j}$ that converges weakly in H^1 to some spherically symmetric $\widehat{\phi}_0(\mathbf{x}) = \widehat{f}_0(|\mathbf{x}|)$. Correspondingly, $\phi^j(\mathbf{x})$ converges weakly in $H^1(B_R)$ to $\phi_0(\mathbf{x}) = f_0(|\mathbf{x}|)$. The important point is that the term $-\int v_- |\phi^j|^2$ is weakly continuous while the term $\int v_+ |\phi^j|^2$ is weakly lower continuous [LLo]. We also note that $f_0(R) = 1$ since the functions $\widehat{\phi^j}$ are identically equal to 1 for $R < |\mathbf{x}| < R + 1$ and the limit $\widehat{\phi}_0$ is continuous away from the origin since it is spherically symmetric and in H^1.

Thus, the limit function ϕ_0 is a minimizer for $\mathcal{E}[\phi]$ under the condition $\phi = 1$ on S_R. Since it is a minimizer, it must be harmonic in $\mathcal{A}$, so (C.7) is true. Eq. (C.6) is standard and is obtained by replacing ϕ_0 by $\phi_0 + \delta\psi$, where ψ is any infinitely differentiable function that is zero for $|\mathbf{x}| \geq R$. The first variation in δ gives (C.6).

Eq. (C.8) is obtained by using integration by parts to compute $\mathcal{E}_R[\phi_0]$.

The uniqueness of the minimizer can be proved in two ways. One way is to note that if $\phi_0 \neq \psi_0$ are two minimizers then, by the convexity noted above, $\mathcal{E}_R[\sqrt{\phi_0^2 + \psi_0^2}] < \mathcal{E}_R[\phi_0] + \mathcal{E}_R[\psi_0]$. The second way is to notice that all minimizers satisfy (C.6), which is a linear, ordinary differential equation for f_0 on $(0, R)$ since all minimizers are spherically symmetric, as we noted. But the solution of such equations, given the value at the end points, is unique. □

We thus see that if the Schrödinger operator on $\mathbb{R}^n$ with potential $\frac{1}{2}v(\mathbf{x})$ has no negative energy bound state then the scattering length in (C.7) is well defined by a variational principle. Our next task is to find some properties of the minimizer ϕ_0. For this purpose we shall henceforth assume that v is *nonnegative*, which guarantees (C.3), of course.

Lemma C.2. *If v is nonnegative then for all $0 < r \le R$ the minimizer $\phi_0(\mathbf{x}) = f_0(|\mathbf{x}|)$ satisfies*

A)

$$f_0(r) \ge f_0^{\mathrm{asymp}}(r), \qquad (C.12)$$

where f_0^{asymp} is given in (C.7)

B) $f_0(r)$ is a monotonically nondecreasing function of r.

C) If $v(r) \ge \tilde{v}(r) \ge 0$ for all r then the corresponding minimizers satisfy $f_0(r) \le \tilde{f}_0(r)$ for all $r < R$. Hence, $a > \tilde{a} \ge 0$.

Proof. Let us define $f_0^{\mathrm{asymp}}(r)$ for *all* $0 < r < \infty$ by (C.7), and let us extend $f_0(r)$ to all $0 < r < \infty$ by setting $f_0(r) = f_0^{\mathrm{asymp}}(r)$ when $r \ge R$.

To prove A) Note that $-\Delta\phi_0 = -\frac{1}{2}v\phi_0$, which implies that ϕ_0 is subharmonic (we use $v \ge 0$ and $\phi_0 \ge 0$, by Theorem C.1). Set $h_\varepsilon(r) = f_0(r) - (1+\varepsilon)f_0^{\mathrm{asymp}}(r)$ with $\varepsilon > 0$ and small. Obviously, $\mathbf{x} \mapsto h_\varepsilon(|\mathbf{x}|)$ is subharmonic on the open set $\{\mathbf{x} : 0 < |\mathbf{x}| < \infty\}$ because $f_0^{\mathrm{asymp}}(|\mathbf{x}|)$ is harmonic there. Clearly, $h_\varepsilon \to -\infty$ as $r \to \infty$ and $h_\varepsilon(R) = -\varepsilon$. Suppose that (C.12) is false at some radius $\rho < R$ and that $h_0(\rho) = -c < 0$. In the annulus $\rho < r < \infty$, $h_\varepsilon(r)$ has its maximum on the boundary, i.e., either at ρ or at ∞ (since $h(|\mathbf{x}|)$ is subharmonic in $\mathbf{x}$). By choosing ε sufficiently small and positive we can have that $h_\varepsilon(\rho) < -2\varepsilon$ and this contradicts the fact that the maximum (which is at least $-\varepsilon$) is on the boundary.

B) is proved by noting (by subharmonicity again) that the maximum of f_0 in $(0, r)$ occurs on the boundary, i.e., $f_0(r) \ge f_0(r')$ for any $r' < r$.

C) is proved by studying the function $g = f_0 - \tilde{f}_0$. Since f_0 and $\tilde{f}_0$ are continuous, the falsity of C) implies the existence some open subset, $\Omega \subset B_R$ on which $g(|\mathbf{x}|) > 0$. On Ω we have that $g(|\mathbf{x}|)$ is subharmonic (because $vf_0 > \tilde{v}\tilde{f}_0$). Hence, its maximum occurs on the boundary, but $g = 0$ there. This contradicts $g(|\mathbf{x}|) > 0$ on Ω. □

Remark about infinite range potentials: If $v(r)$ is infinite range and nonnegative it is easy to extend the definition of the scattering length under the assumptions:

    1) $v(r) \geq 0$ for all r and

    2) For some R_1 we have $\int_{R_1}^{\infty} v(r) r^{n-1} \, dr < \infty$.

    If we cut off the potential at some point $R_0 > R_1$ (i.e., set $v(r) = 0$ for $r > R_0$) then the scattering length is well defined but it will depend on R_0, of course. Denote it by $a(R_0)$. By part C of Lemma (C.2), $a(R_0)$ is an increasing function of R_0. However, the bounds (C.9) and (C.10) and assumption 2) above guarantee that $a(R_0)$ is bounded above. (More precisely, we need a simple modification of (C.9) and (C.10) to the potential $\widehat{v}(r) \equiv \infty$ for $r \leq R_1$ and $\widehat{v}(r) \equiv v(r)$ for $r > R_1$. This is accomplished by replacing the 'trial function' $f(x) = 1$ by a smooth radial function that equals 0 for $r < R_1$ and equals 1 for $r > R_2$ for some $R_2 > R_1$.) Thus, a is well defined by

$$a = \lim_{R_0 \to \infty} a(R_0) \, . \tag{C.13}$$

Appendix D

c-Number Substitutions and Gauge Symmetry Breaking

In this appendix, which is a slightly extended version of [LSeY8], we give a rigorous justification of part of the Bogoliubov approximation, discussed in Appendix A – namely the replacement of bosonic creation and annihilation operators by c-numbers. We also discuss the relation between BEC and spontaneous breaking of gauge symmetry which was mentioned briefly in the introduction.

One of the key developments in the theory of the Bose gas, especially the theory of the low density gases currently at the forefront of experiment, is Bogoliubov's 1947 analysis [Bo] of the many-body Hamiltonian by means of a c-number substitution for the most relevant operators in the problem, the zero-momentum mode operators, namely $a_0 \to z$, $a_0^* \to z^*$. Naturally, the appropriate value of z has to be determined by some sort of consistency or variational principle, which might be complicated, but the concern, expressed by many authors over the years, is whether this sort of substitution is legitimate, i.e., error free. We address this latter problem here and show, by a simple but rigorous analysis, that it is so under very general circumstances.

The rigorous justification for this substitution, as far as calculating the pressure is concerned, was done in a classic paper of Ginibre [Gin] in 1968, but it does not seem to have percolated into the general theory community. In textbooks it is often said, for instance, that it is tied to the imputed 'fact' that the expectation value of the number operator $n_0 = a_0^* a_0$ is of order V = volume. (This was the argument in [Bo]). That is, Bose-Einstein condensation (BEC) justifies the substitution. As Ginibre pointed out, however, BEC has nothing to do with it. The z substitution still gives the right answer even if n_0 is small (but it is a useful calculational tool only if n_0 is macroscopic). Thus, despite [Gin] and the thorough review of these matters in [ZB], there is some confusion in the literature and clarification could be useful.

In this appendix we do three things. 1.) We show how Ginibre's result can be easily obtained in a few simple lines. While he used coherent states, he did not use the Berezin-Lieb inequality [Be, L4, Si2], derived later, which efficiently gives upper bounds. This inequality gives explicit error bounds which, typically, are only order one compared to the total free energy or pressure times volume, which are order N = particle number.

2.) This allows us to go beyond [Gin] and make c-number substitutions for *many* **k**-modes at once, provided the number of modes is lower order than N.

3.) We show how the optimum value of z yields, in fact, the expectation value $\langle n_0 \rangle$ in the true state when a gauge breaking term is added to the Hamiltonian. More precisely, in the thermodynamic limit (TL) the $|z|^2$ that maximizes the partition function equals $|\langle a_0 \rangle|^2$ and this equals $\langle n_0 \rangle$, which is the amount of condensation — a point that was not addressed in full generality in previous work [Gin, ZB, BSP, AVZ]. The second of these equalities has previously only been treated under some additional assumptions [FPV] or for some simplified models [ZB, Su].[1]

While we work here at positive temperature $k_{\mathrm{B}}T = 1/\beta$, our methods also work for the ground state (and are even simpler in that case). To keep this note short and, hopefully, readable, we will be a bit sketchy in places but there is no difficulty filling in the details.

The use of coherent states [KS, ZFG] to give accurate upper and lower bounds to energies, and thence to expectation values, is effective in a wide variety of problems [L6], e.g., quantum spin systems in the large S limit [L4], the Dicke model [HL], the strong coupling polaron [LT], and the proof that Thomas-Fermi theory is exact in the large atom limit [L5, Th]. For concreteness and relevance, we concentrate on the Bose gas problem here, and we discuss only the total, correct Hamiltonian. Nevertheless, the same conclusions hold also for variants, such as Bogoliubov's truncated Hamiltonian (the "weakly imperfect Bose gas" [ZB, Bo]) or other modifications, provided we are in the stability regime (i.e., the regime in which the models make sense). We are not claiming that any particular

---

[1] We note that recently A. Sütő presented a different proof of item 3 [Su2].

approximation is valid. That is a completely different story that has to be decided independently. The method can also be modified to incorporate inhomogeneous systems. The message is the same in all cases, namely that the z substitution causes no errors (in the TL), even if there is no BEC, whenever it is applied to physically stable systems. Conversely, if the system is stable after the z substitution then so is the original one.

We start with the well-known Hamiltonian for bosons in a large box of volume V, expressed in terms of the second-quantized creation and annihilation operators $a_{\mathbf{k}}, a_{\mathbf{k}}^*$ satisfying the canonical commutation relations,

$$H = \sum_{\mathbf{k}} k^2 a_{\mathbf{k}}^* a_{\mathbf{k}} + \frac{1}{2V} \sum_{\mathbf{k},\mathbf{p},\mathbf{q}} \nu(\mathbf{p}) a_{\mathbf{k}+\mathbf{p}}^* a_{\mathbf{q}-\mathbf{p}}^* a_{\mathbf{k}} a_{\mathbf{q}}, \tag{D.1}$$

(with $\hbar = 2m = 1$). Here, ν is the Fourier transform of the two-body potential $v(\mathbf{x})$. We assume that there is a bound on the Fourier coefficients $|\nu(\mathbf{k})| \leq \varphi < \infty$.

The case of hard core potentials can be taken care of in the following way. First cut off the hard core potential v at a height 10^{12} eV. It is easy to prove, by standard methods, that this cutoff will have a negligible effect on the exact answer. After the cutoff φ will be about 10^{12} eV Å^3, and according to what we prove below, this substitution will affect the chemical potential only by about φ/V, which is truly negligible when $V = 10^{23}$ Å^3.

If we replace the operator $a_{\mathbf{0}}$ by a complex number z and $a_{\mathbf{0}}^*$ by z^* everywhere in H we obtain a Hamiltonian $H'(z)$ that acts on the Fock-space of all the modes other than the $a_{\mathbf{0}}$ mode. Unfortunately, $H'(z)$ does not commute with the particle number $N^> \equiv \sum_{\mathbf{k}\neq 0} a_{\mathbf{k}}^* a_{\mathbf{k}}$. It is convenient, therefore, to work in the grand-canonical ensemble and consider $H_\mu = H - \mu N = H - \mu(a_{\mathbf{0}}^* a_{\mathbf{0}} + N^>)$ and, correspondingly, $H'_\mu(z) = H'(z) - \mu(|z|^2 + N^>)$.

The partition functions are given by

$$e^{\beta V p(\mu)} \equiv \Xi(\mu) = \mathrm{tr}_{\mathcal{H}} \exp[-\beta H_\mu] \tag{D.2}$$

$$e^{\beta V p'(\mu)} \equiv \Xi'(\mu) = \int d^2 z \, \mathrm{tr}_{\mathcal{H}'} \exp[-\beta H'_\mu(z)] \tag{D.3}$$

where $\mathcal{H}$ is the full Hilbert (Fock) space, $\mathcal{H}'$ is the Fock space without the $a_{\mathbf{0}}$ mode, and $d^2 z \equiv \pi^{-1} dx dy$ with $z = x + iy$. The functions $p(\mu)$ and $p'(\mu)$ are the corresponding finite volume pressures.

The pressure $p(\mu)$ has a finite TL for all $\mu < \mu_{\text{critical}}$, and it is a convex function of μ. For the non-interacting gas, $\mu_{\text{critical}} = 0$, but for any *realistic* system $\mu_{\text{critical}} = +\infty$. In any case, we assume $\mu < \mu_{\text{critical}}$, in which case both the pressure and the density are finite.

Let $|z\rangle = \exp\{-|z|^2/2 + za_0^*\}|0\rangle$ be the coherent state vector in the a_0 Fock space and let $\Pi(z) = |z\rangle\langle z|$ be the projector onto this vector. There are six relevant operators containing a_0 in H_μ, which have the following expectation values [KS] (called *lower symbols*)

$$\langle z|a_0|z\rangle = z, \qquad \langle z|a_0 a_0|z\rangle = z^2, \qquad \langle z|a_0^* a_0|z\rangle = |z|^2$$

$$\langle z|a_0^*|z\rangle = z^*, \qquad \langle z|a_0^* a_0^*|z\rangle = z^{*2}, \qquad \langle z|a_0^* a_0^* a_0 a_0|z\rangle = |z|^4.$$

Each also has an *upper symbol*, which is a function of z (call it $u(z)$ generically) such that an operator F is represented as $F = \int d^2 z\, u(z)\Pi(z)$. These symbols are

$$a_0 \rightarrow z, \qquad a_0 a_0 \rightarrow z^2, \qquad a_0^* a_0 \rightarrow |z|^2 - 1$$

$$a_0^* \rightarrow z^*, \qquad a_0^* a_0^* \rightarrow z^{*2}, \qquad a_0^* a_0^* a_0 a_0 \rightarrow |z|^4 - 4|z|^2 + 2.$$

It will be noted that the operator $H'_\mu(z)$, defined above, is obtained from H_μ by substituting the lower symbols for the six operators. If we substitute the upper symbols instead into H_μ we obtain a slightly different operator, which we write as $H''_\mu(z) = H'_\mu(z) + \delta_\mu(z)$ with

$$\delta_\mu(z) = \mu + \frac{1}{2V}\left[(-4|z|^2 + 2)\nu(0) - \sum_{k \neq 0} a_k^* a_k\big(2\nu(0) + \nu(k) + \nu(-k)\big)\right]. \quad (D.4)$$

The next step is to mention two inequalities, of which the first is

$$\Xi(\mu) \geq \Xi'(\mu). \qquad (D.5)$$

This is a consequence of the following two facts: The completeness property of coherent states, $\int d^2 z\, \Pi(z) = $ Identity, and

$$\langle z \otimes \phi|e^{-\beta H_\mu}|z \otimes \phi\rangle \geq e^{-\beta\langle z \otimes \phi|H_\mu|z \otimes \phi\rangle} = e^{-\beta\langle \phi|H'_\mu(z)|\phi\rangle}, \qquad (D.6)$$

where ϕ is any normalized vector in $\mathcal{H}'$. This is Jensen's inequality for the expectation value of a convex function (like the exponential function) of an operator.

To prove (D.5) we take ϕ in (D.6) to be one of the normalized eigenvectors of $H'_\mu(z)$, in which case $\exp\{\langle\phi| - \beta H'_\mu(z)|\phi\rangle\} = \langle\phi|\exp\{-\beta H'_\mu(z)\}|\phi\rangle$. We then sum over all such eigenvectors (for a fixed z) and integrate over z. The left side is then $\Xi(\mu)$, while the right side is $\Xi'(\mu)$.

The second inequality [Be, L4, Si2] is

$$\Xi(\mu) \leq \Xi''(\mu) \equiv e^{\beta V p''(\mu)}, \qquad (D.7)$$

where $\Xi''(\mu)$ is similar to $\Xi'(\mu)$ except that $H'_\mu(z)$ is replaced by $H''_\mu(z)$. Its proof is the following. Let $|\Phi_j\rangle \in \mathcal{H}$ denote the complete set of normalized eigenfunctions of H_μ. The partial inner product $|\Psi_j(z)\rangle = \langle z|\Phi_j\rangle$ is a vector in $\mathcal{H}'$ whose square norm, given by $c_j(z) = \langle\Psi_j(z)|\Psi_j(z)\rangle_{\mathcal{H}'}$, satisfies $\int d^2z\, c_j(z) = 1$. By using the upper symbols, we can write $\langle\Phi_j|H_\mu|\Phi_j\rangle = \int d^2z\,\langle\Psi_j(z)|H''_\mu(z)|\Psi_j(z)\rangle = \int d^2z\,\langle\Psi'_j(z)|H''_\mu(z)|\Psi'_j(z)\rangle c_j(z)$, where $|\Psi'_j(z)\rangle$ is the normalized vector $c_j(z)^{-1/2}\Psi_j(z)$. To compute the trace, we can exponentiate this to write $\Xi(\mu)$ as

$$\sum_j \exp\left\{-\beta\int d^2z\,c_j(z)\langle\Psi'_j(z)|H''_\mu(z)|\Psi'_j(z)\rangle\right\}.$$

Using Jensen's inequality twice, once for functions and once for expectations as in (D.6), $\Xi(\mu)$ is less than

$$\sum_j \int d^2z\,c_j(z)\exp\left\{\langle\Psi'_j(z)| - \beta H''_\mu(z)|\Psi'_j(z)\rangle\right\}$$

$$\leq \sum_j \int d^2z\,c_j(z)\langle\Psi'_j(z)|\exp\left\{-\beta H''_\mu(z)\right\}|\Psi'_j(z)\rangle.$$

Since $\operatorname{tr}\Pi(z) = 1$, the last expression can be rewritten

$$\int d^2z \sum_j \langle\Phi_j|\Pi(z)\otimes\exp\left\{-\beta H''_\mu(z)\right\}|\Phi_j\rangle = \Xi''(\mu).$$

Thus, we have that

$$\Xi'(\mu) \leq \Xi(\mu) \leq \Xi''(\mu). \tag{D.8}$$

The next step is to try to relate $\Xi''(\mu)$ to $\Xi'(\mu)$. To this end we have to bound $\delta_\mu(z)$ in (D.4). This is easily done in terms of the total number operator whose lower symbol is $N'(z) = |z|^2 + \sum_{k\neq 0} a^*_k a_k$. In terms of the bound φ on $\nu(\mathbf{p})$

$$|\delta_\mu(z)| \leq 2\varphi(N'(z) + \tfrac{1}{2})/V + |\mu|. \tag{D.9}$$

Consequently, $\Xi''(\mu)$ and $\Xi'(\mu)$ are related by the inequality

$$\Xi''(\mu) \leq \Xi'(\mu + 2\varphi/V)e^{\beta(|\mu|+\varphi/V)}. \tag{D.10}$$

Equality of the pressures $p(\mu)$, $p'(\mu)$ and $p''(\mu)$ in the TL follows from (D.8) and (D.10).

Closely related to this point is the question of relating $\Xi(\mu)$ to the maximum value of the integrand in (D.3), which is $\max_z \operatorname{tr}_{\mathcal{H}'}\exp[-\beta H'_\mu(z)] \equiv e^{\beta V p^{\max}}$. This latter quantity is often used in discussions of the z substitution problem, e.g., in refs. [Gin, ZB]. One direction is not hard. It is the inequality (used in ref. [Gin])

$$\Xi(\mu) \geq \max_z \operatorname{tr}_{\mathcal{H}'} \exp[-\beta H'_\mu(z)], \tag{D.11}$$

and the proof is the same as the proof of (D.5), except that this time we replace the completeness relation for the coherent states by the simple inequality Identity $\geq$ $\Pi(z)$ for any fixed number z.

For the other direction, split the integral defining $\Xi''(\mu)$ into a part where $|z|^2 < \xi$ and $|z|^2 \geq \xi$. Thus,

$$\Xi''(\mu) \leq \xi \max_z \operatorname{tr}_{\mathcal{H}'} \exp[-\beta H''_\mu(z)] + \frac{1}{\xi} \int_{|z|^2 \geq \xi} d^2z \, |z|^2 \, \operatorname{tr}_{\mathcal{H}'} \exp[-\beta H''_\mu(z)]. \tag{D.12}$$

Dropping the condition $|z|^2 \geq \xi$ in the last integral and using $|z|^2 \leq N'(z) = N''(z) + 1$, we see that the last term on the right side of (D.12) is bounded above by $\xi^{-1}\Xi''(\mu)[V\rho''(\mu) + 1]$, where $\rho''(\mu)$ denotes the density in the H''_μ ensemble. Optimizing over ξ leads to

$$\Xi''(\mu) \leq 2[V\rho''(\mu) + 1] \max_z \operatorname{tr}_{\mathcal{H}'} \exp[-\beta H''_\mu(z)]. \tag{D.13}$$

Note that $\rho''(\mu)$ is order one, since $p''(\mu)$ and $p(\mu)$ agree in the TL (and are convex in μ), and we assumed that the density in the original ensemble is finite. By (D.9), $H''_\mu \geq H'_{\mu+2\varphi/V} - |\mu| - \varphi/V$, and it follows from (D.7), (D.13) and (D.11) that $p^{\max}$ agrees with the true pressure p in the TL. Their difference, in fact, is at most $O(\ln V/V)$. This is the result obtained by Ginibre in [Gin] by more complicated arguments, under the assumption of superstability of the interaction, and without the explicit error estimates obtained here.

To summarize the situation so far, we have four expressions for the grand-canonical pressure. They are all equal in the TL limit,

$$p(\mu) = p'(\mu) = p''(\mu) = p^{\max}(\mu) \tag{D.14}$$

when μ is not a point at which the density can be infinite.

Our second main point is that not only is the z substitution valid for a_0 but it can also be done for many modes simultaneously. As long as the number of modes treated in this way is much smaller than N the effect on the pressure will be negligible. Each such substitution will result in an error in the chemical potential that is order φ/V. The proof of this fact just imitates what was done above for one mode. Translation invariance is not important here; one can replace any mode such as $\sum_\mathbf{k} g_\mathbf{k} a_\mathbf{k}$ by a c-number, which can be useful for inhomogeneous systems.

A more delicate point is our third one, and it requires, first, a discussion of the meaning of 'condensate fraction' that goes beyond what is usually mentioned in textbooks, but which was brought out in [Bo, Gri, Ro]. The 'natural' idea would be to consider $V^{-1}\langle n_0\rangle$. This, however, need not be a reliable measure of the condensate fraction for the following reason. If we expand $\exp\{-\beta H\}$ in eigenfunctions of the number operator n_0 we would have $\langle n_0\rangle = \sum_n n\gamma(n)$, where $\gamma(n)$ is the probability that $n_0 = n$. One would like to think that $\gamma(n)$ is sharply peaked at some maximum n value, but we do not know if this is the case. $\gamma(n)$ could be flat, up to the maximum value or, worse, it could have a maximum at $n = 0$. Recall that precisely this happens for the Heisenberg quantum ferromagnet [Gri]; by virtue of conservation of total spin angular momentum, the distribution of values of the z-component of the total spin, S^z, is a strictly decreasing function of $|S^z|$. Even if it were flat, the expected value of S^z would be half of the spontaneous magnetization that one gets by applying a weak magnetic field.

With this example in mind, we see that the only physically reliable quantity is $\lim_{\lambda\to 0} \lim_{V\to\infty} V^{-1}\langle n_0\rangle_{\mu,\lambda}$, where the expectation is now with respect to a Hamiltonian $H_{\mu,\lambda} = H_\mu + \sqrt{V}(\lambda a_0 + \lambda^* a_0^*)$ [Bo]. Without loss of generality, we assume λ to be real. We will show that for almost every λ, the density $\gamma(V\rho_0)$ converges in the TL to a δ-function at the point $\hat{\rho}_0 = \lim_{V\to\infty} |z_{\max}|^2/V$, where $z_{\max}$ maximizes the partition function $\mathrm{tr}_{\mathcal{H}'} \exp\{-\beta H'_{\mu,\lambda}(z)\}$. That is,

$$\lim_{V\to\infty} \frac{1}{V}\langle n_0\rangle_{\mu,\lambda} = \lim_{V\to\infty} \frac{1}{V}|\langle a_0\rangle_{\mu,\lambda}|^2 = \lim_{V\to\infty} \frac{1}{V}|z_{\max}|^2. \qquad (D.15)$$

This holds for those λ where the pressure in the TL is differentiable; since $p(\mu,\lambda)$ is convex (upwards) in λ this is true almost everywhere. The right and left derivatives exist for every λ and hence $\lim_{\lambda\to 0+} \lim_{V\to\infty} V^{-1}|\langle a_0\rangle_{\mu,\lambda}|^2$ exists.

The expectation values $\langle n_0\rangle_{\mu,\lambda}$ and $\langle a_0\rangle_{\mu,\lambda}$ are obtained by integrating $(|z|^2 - 1)$ and z, respectively, with the weight $W_{\mu,\lambda}(z)$, given by $W_{\mu,\lambda}(z) \equiv \Xi(\mu,\lambda)^{-1}\mathrm{tr}_{\mathcal{H}'}\langle z| \exp\{-\beta H_{\mu,\lambda}\}|z\rangle$. We will show that this weight converges to a δ-function at $z_{\max}$ in the TL, implying (D.15). If we could replace $W_{\mu,\lambda}(z)$ by $W_{\mu,0}(z)e^{-\beta\lambda\sqrt{V}(z+z^*)}$, this would follow from Griffiths' argument [Gri] (see also [DLS, Sect. 1]). Because $[H, a_0] \neq 0$, $W_{\mu,\lambda}$ is not of this product form. However, the weight for $\Xi''(\mu,\lambda)$, which is $W''_{\mu,\lambda}(z) \equiv \Xi''(\mu,\lambda)^{-1}\mathrm{tr}_{\mathcal{H}'}\exp\{-\beta H''_{\mu,\lambda}(z)\}$, does have the right form. In the following we shall show that the two weights are equal apart from negligible errors.

Equality (D.14) holds also for all λ, i.e., $p(\mu, \lambda) = p''(\mu, \lambda) = p^{\max}(\mu, \lambda)$ in the TL. In fact, since the upper and lower symbols agree for a_0 and a_0^*, the error estimates above remain unchanged. (Note that since $\sqrt{V}|a_0 + a_0^*| \leq \delta(N + \frac{1}{2}) + V/\delta$ for any $\delta > 0$, $p(\mu, \lambda)$ is finite for all λ if it is finite for $\lambda = 0$ in a small interval around μ.) At any point of differentiability with respect to λ, Griffiths' theorem [Gri] (see [DLS, Cor. 1.1]), applied to the partition function $\Xi''(\lambda, \mu)$, implies that $W_{\mu,\lambda}''(\zeta\sqrt{V})$ converges to a δ-function at some point $\widehat{\zeta}$ on the real axis as $V \to \infty$. (The original Griffiths argument can easily be extended to two variables, as we have here. Because of radial symmetry, the derivative of the pressure with respect to $\operatorname{Im}\lambda$ is zero at any non-zero real λ.) Moreover, by comparing the derivatives of p'' and $p^{\max}$ we see that $\widehat{\zeta} = \lim_{V\to\infty} z_{\max}/\sqrt{V}$, since $z_{\max}/\sqrt{V}$ is contained in the interval between the left and right derivatives of $p^{\max}(\mu, \lambda)$ with respect to λ.

We shall now show that the same is true for $W_{\mu,\lambda}$. To this end, we add another term to the Hamiltonian, namely $\varepsilon F \equiv \varepsilon V \int d^2z\, \Pi(z) f(zV^{-1/2})$, with ε and f real. If $f(\zeta)$ is a nice function of two real variables with bounded second derivatives, it is then easy to see that the upper and lower symbols of F differ only by a term of order 1. Namely, for some $C > 0$ independent of z_0 and V,

$$\left| V \int d^2z\, |\langle z|z_0\rangle|^2 \left(f(zV^{-1/2}) - f(z_0 V^{-1/2}) \right) \right| \leq C.$$

Hence, in particular, $p(\mu, \lambda, \varepsilon) = p''(\mu, \lambda, \varepsilon)$ in the TL. Moreover, if $f(\zeta) = 0$ for $|\zeta - \widehat{\zeta}| \leq \delta$, then the pressure is independent of ε for $|\varepsilon|$ small enough (depending only on δ). This can be seen as follows. We have

$$p''(\mu, \lambda, \varepsilon) - p''(\mu, \lambda, 0) = \frac{1}{\beta V} \ln \left\langle e^{-\beta\varepsilon V f(zV^{-1/2})} \right\rangle, \tag{D.16}$$

where the last expectation is in the H_μ'' ensemble at $\varepsilon = 0$. The corresponding distribution is exponentially localized at $z/\sqrt{V} = \widehat{\zeta}$ [Gri, DLS], and therefore the right side of (D.16) goes to zero in the TL for small enough ε. In particular, the ε-derivative of the TL pressure at $\varepsilon = 0$ is zero. By convexity in ε, this implies that the derivative of p at finite volume, given by $V^{-1}\langle F\rangle_{\mu,\lambda} = \int d^2z\, f(zV^{-1/2})W_{\mu,\lambda}(z)$, goes to zero in the TL. Since f was arbitrary, $V \int_{|\zeta-\widehat{\zeta}|\geq\delta} d^2\zeta\, W_{\mu,\lambda}(\zeta\sqrt{V}) \to 0$ as $V \to \infty$. This holds for all $\delta > 0$, and therefore proves the statement.

Our method also applies to the case when the pressure is not differentiable in λ (which is the case at $\lambda = 0$ in the presence of BEC). In this case, the resulting weights $W_{\mu,\lambda}$ and $W_{\mu,\lambda}''$ need not be δ-functions, but Griffiths' method [Gri, DLS] implies that they are, for $\lambda \neq 0$, supported on the real axis between the right and left derivative of p and, for $\lambda = 0$, on a disc (due to the gauge symmetry) with radius determined by the right derivative at $\lambda = 0$. Convexity of the pressure as a function of λ thus implies that in the TL the supports of the weights $W_{\mu,\lambda}$ and $W''_{\mu,\lambda}$ for $\lambda \neq 0$ lie outside of this disc. Hence $\langle n_0 \rangle_\lambda$ is monotone increasing in λ in the TL. In combination with (D.15) this implies in particular that

$$\lim_{V\to\infty} \frac{1}{V} \langle n_0 \rangle_{\mu,\lambda=0} \leq \lim_{\lambda\to 0} \lim_{V\to\infty} \frac{1}{V} |\langle a_0 \rangle_{\mu,\lambda}|^2 , \qquad (D.17)$$

a fact which is intuitively clear but has, to the best of our knowledge, not been proved so far [Ro] in this generality. In fact, the only hypothesis entering our analysis, apart from the bound φ on the potential, is the existence of the TL of the pressure and the density.

We note that by Eq. (D.17) spontaneous symmetry breaking (in the sense that the right side of (D.17) is not zero) always takes place whenever there is BEC is the usual sense, i.e., without explicit gauge breaking (meaning that the left side of (D.17) is non-zero). Conversely, by Eq. (D.15) spontaneous symmetry breaking is equivalent to BEC in the sense of 'quasi-averages' [Bo2], i.e.,

$$\lim_{\lambda\to 0} \lim_{V\to\infty} \frac{1}{V} \langle n_0 \rangle_{\mu,\lambda} > 0 . \qquad (D.18)$$

Note, however, that a non-vanishing of the right side of (D.17) does not *a priori* imply a non-vanishing of the left side. I.e., it is *a priori* possible that BEC only shows up after introducing an explicit gauge-breaking term to the Hamiltonian. While it is expected on physical grounds that positivity of the right side of (D.17) implies positivity of the left side, a rigorous proof is lacking, so far. In the example of the Heisenberg magnet quoted above, equality in (D.17) does not generally hold, but still both sides are non-vanishing in the same parameter regime.

To illustrate what could arise mathematically, in principle, consider a weight function of the form

$$W''_{\mu,\lambda=0}(\sqrt{V}\zeta) \equiv w_V(\zeta) = \begin{cases} V^2 - V + 1/V & \text{for } |\zeta| \leq 1/V \\ 1/V & \text{for } 1/V \leq |\zeta| \leq 1 \\ 0 & \text{for } |z| > 1 . \end{cases} \qquad (D.19)$$

This distribution converges for $V \to \infty$ to a δ-function at $\zeta = 0$, and consequently there is no BEC at $\lambda = 0$. On the other hand, it is easy to see that the weight function $w_V(\zeta)e^{-\beta\lambda V\zeta}$ (with an appropriate normalization factor) converges, for any $\lambda > 0$, to a δ-function at $\zeta = -1$ as $V \to \infty$, and hence there is spontaneous symmetry breaking. The open problem for the mathematician is to prove that examples like (D.19) do not occur in realistic bosonic systems.

Bibliography

[ALSSY] M. Aizenman, E.H. Lieb, R. Seiringer, J.P. Solovej, and J. Yngvason, *Bose-Einstein quantum phase transition in an optical lattice model*, Phys. Rev. A **70**, 023612-1–12 (2004).

[ALSSY2] M. Aizenman, E.H. Lieb, R. Seiringer, J.P. Solovej, and J. Yngvason, *Bose-Einstein Condensation as a Quantum Phase Transition in an Optical Lattice*, arXiv:cond-mat/0412034, to appear in the proceedings of QMath9, Giens, France, Sept. 12–16, 2004.

[AN] M. Aizenman, B. Nachtergaele, *Geometric Aspects of Quantum Spin States*, Commun. Math. Phys. **164**, 17–63 (1994).

[AVZ] N. Angelescu, A. Verbeure, V.A. Zagrebnov, *On Bogoliubov's model of superfluidity*, J. Phys. A: Math. Gen. **25**, 3473-3491 (1992); *Superfluidity III*, J. Phys. A: Math. Gen. **30**, 4895-4913 (1997).

[AA] E. Altman, A. Auerbach. *Oscillating Superfluidity of Bosons in Optical Lattices*, Phys. Rev. Lett. **89**, 250404 (2002).

[AG] G.E. Astrakharchik and S. Giorgini, *Quantum Monte Carlo study of the three- to one-dimensional crossover for a trapped Bose gas*, Phys. Rev. A **66**, 053614-1–6 (2002).

[ABCG] G.E. Astrakharchik, J. Boronat, J. Casulleras, and S. Giorgini, *Superfluidity versus Bose-Einstein condensation in a Bose gas with disorder*, Phys. Rev. A **66**, 023603 (2002).

[Ba] B. Baumgartner, *The Existence of Many-particle Bound States Despite a Pair Interaction with Positive Scattering Length*, J. Phys. A **30**, L741–L747 (1997).

[Bm]   G. Baym, in: *Mathematical Methods in Solid State and Superfluid The-
          ory*, Scottish Univ. Summer School of Physics, Oliver and Boyd, Ed-
          inburgh (1969).

[Be]   F.A. Berezin, Izv. Akad. Nauk, ser. mat., **36** (No. 5) (1972); English
          translation: USSR Izv. **6** (No. 5) (1972). F.A. Berezin, *General concept
          of quantization*, Commun. Math. Phys. **40**, 153–174 (1975).

[BL]   M. van den Berg, J.T. Lewis, *On generalized condensation in the free
          Bose gas*, Physica A **110**, 550–564 (1982).

[BLP]   M. van den Berg, J.T. Lewis, J.V. Pulè, *A general theory of Bose-
          Einstein condensation*, Helv. Phys. Acta **59**, 1271–1288 (1986).

[Bl]   D. Blume, *Fermionization of a bosonic gas under highly elongated con-
          finement: A diffusion quantum Monte Carlo study*, Phys. Rev. A **66**,
          053613-1–8 (2002).

[Bo]   N.N. Bogoliubov, *On the theory of superfluidity*, Izv. Akad. Nauk
          USSR, **11**, 77 (1947). Eng. Trans. J. Phys. (USSR), **11**, 23 (1947). See
          also *Lectures on quantum statistics*, vol. 1, Gordon and Breach (1967).

[Bo2]   N.N. Bogoliubov, *Lectures on Quantum Statistics: Quasi-Averages*,
          Gordon and Breach (1970).

[Bo3]   N.N. Bogoliubov, *Energy levels of the imperfect Bose-Einstein gas*,
          Bull. Moscow State Univ. **7**, 43–56 (1947).

[BZ]   N.N. Bogoliubov, D.N. Zubarev, *Wave function of the ground state of
          interacting Bose-particles*, Sov. Phys.-JETP **1**, 83 (1955).

[BZT]   N.N. Bogoliubov, D.N. Zubarev, Y.A. Tserkovnikov, Soviet Phys. Dok-
          lady **2**, 535 (1957).

[BBD]   K. Bongs, S. Burger, S. Dettmer, D. Hellweg, J. Artl, W. Ertmer, and
          K. Sengstok, *Waveguides for Bose-Einstein condensates*, Phys. Rev.
          A, **63**, 031602 (2001).

[B]   S.N. Bose, *Plancks Gesetz und Lichtquantenhypothese*, Z. Phys. **26**,
          178–181 (1924).

[BSP]   E. Buffet, Ph. de Smedt, J.V. Pulè, *The condensate equation for some
          Bose systems*, J. Phys. A **16**, 4307 (1983).

[C]   Y. Castin, *Bose-Einstein condensates in atomic gases: simple theoret-ical results*, in "Coherent atomic matter waves", Lecture Notes of Les Houches Summer School, pp. 1–136, edited by R. Kaiser, C. Westbrook, and F. David, EDP Sciences and Springer-Verlag (2001).

[CS1]   A.Y. Cherny, A.A. Shanenko, *Dilute Bose gas in two dimensions: Den-sity expansions and the Gross-Pitaevskii equation*, Phys. Rev. E **64**, 027105 (2001).

[CS2]   A.Y. Cherny, A.A. Shanenko, *The kinetic and interaction energies of a trapped Bose gas: Beyond the mean field*, Phys. Lett. A **293**, 287 (2002).

[CLY]   J. Conlon, E.H. Lieb, H.-T. Yau, *The $N^{7/5}$ Law for Charged Bosons*, Commun. Math. Phys. **116**, 417–448 (1988).

[CCRCW]   S.L. Cornish, N.R. Claussen, J.L. Roberts, E.A. Cornell, C.E. Wieman, *Stable ^{85}Rb Bose-Einstein Condensates with Widely Tunable Interac-tions*, Phys. Rev. Lett. **85**, 1795–98 (2000).

[DGPS]   F. Dalfovo, S. Giorgini, L.P. Pitaevskii, S. Stringari, *Theory of Bose-Einstein condensation in trapped gases*, Rev. Mod. Phys. **71**, 463–512 (1999).

[Da]   K.K. Das, *Highly anisotropic Bose-Einstein condensates: Crossover to lower dimensionality*, Phys. Rev. A **66**, 053612-1–7 (2002).

[DGW]   K.K. Das, M.D. Girardeau, and E.M. Wright, *Crossover from One to Three Dimensions for a Gas of Hard-Core Bosons*, Phys. Rev. Lett. **89**, 110402-1–4 (2002).

[DODS]   D.B.M. Dickerscheid, D. van Oosten, P.J.H. Denteneer, H.T.C. Stoof, *Ultracold atoms in optical lattices*, Phys. Rev. A **68**, 043623 (2003).

[DLO]   V. Dunjko, V. Lorent, and M. Olshanii, *Bosons in Cigar-Shaped Traps: Thomas-Fermi Regime, Tonks-Girardeau Regime, and In Between*, Phys. Rev. Lett. **86**, 5413–5316 (2001).

[D1]   F.J. Dyson, *Ground-State Energy of a Hard-Sphere Gas*, Phys. Rev. **106**, 20–26 (1957).

[D2]   F.J. Dyson, *Ground State Energy of a Finite System of Charged Par-ticles*, J. Math. Phys. **8**, 1538–1545 (1967).

[DLS]   F.J. Dyson, E.H. Lieb, B. Simon, *Phase Transitions in Quantum Spin Systems with Isotropic and Nonisotropic Interactions*, J. Stat. Phys. **18**, 335–383 (1978).

[E]   A. Einstein, *Quantentheorie des einatomigen idealen Gases*, Sitzber. Kgl. Preuss. Akad. Wiss., 261–267 (1924), and 3–14 (1925).

[FPV]   M. Fannes, J.V. Pulè, V. Verbeure, *On Bose condensation*, Helv. Phys. Acta **55**, 391 (1982).

[FS]   A.L. Fetter and A.A. Svidzinsky, *Vortices in a trapped dilute Bose-Einstein condensate*, J. Phys.: Condens. Matter **13**, R135 (2001).

[FH]   D.S. Fisher, P.C. Hohenberg, *Dilute Bose gas in two dimensions*, Phys. Rev. B **37**, 4936–4943 (1988).

[FWGF]   M.P.A. Fisher, P.B. Weichman, G. Grinstein, D.S. Fisher, *Boson localization and the superfluid-insulator transition*, Phys. Rev. B **40**, 546–570 (1989).

[F]   L.L. Foldy, *Charged Boson Gas*, Phys. Rev. **124**, 649–651 (1961); Errata *ibid* **125**, 2208 (1962).

[FL]   J. Fröhlich, E.H. Lieb, *Phase Transitions in Anisotropic Lattice Spin Systems*, Commun. Math. Phys. **60**, 233–267 (1978).

[FSS]   J. Fröhlich, B. Simon, T. Spencer, *Phase Transitions and Continuous Symmetry Breaking*, Phys. Rev. Lett. **36**, 804 (1976); *Infrared bounds, phase transitions and continuous symmetry breaking*, Commun. Math. Phys. **50**, 79 (1976).

[Ga]   J.J. Garcia-Ripoll, J.I. Cirac, P. Zoller, C. Kollath, U. Schollwoeck, J. von Delft, *Variational ansatz for the superfluid Mott-insulator transition in optical lattices*, Opt. Express **12**, 42 (2004).

[Ge]   G.M. Genkin, *Manipulating the superfluid – Mott insulator transition of a Bose-Einstein condensate in an amplitude-modulated optical lattice*, arXiv:cond-mat/0311589 (2003).

[Gin]   J. Ginibre, *On the asymptotic exactness of the Bogoliubov approximation for many boson systems*, Commun. Math. Phys. **8**, 26–51 (1968).

[GA]   M. Girardeau, R. Arnowitt, *Theory of Many-Boson Systems: Pair Theory*, Phys. Rev. **113**, 755 (1959).

[Gi1]   M. Girardeau, *Weak-Coupling Expansion for the Ground-State Energy of a Many-Boson System*, Phys. Rev. **115**, 1090 (1959).

[Gi2]   M.D. Girardeau, *Relationship between systems of impenetrable bosons and fermions in one dimension*, J. Math. Phys. **1**, 516 (1960).

[Gi3]   M. Girardeau, *Simple and generalized condensation in many-boson systems*, Phys. of Fluids **5**, 1468 (1962).

[Gi4]   M. Girardeau, *Ground State of the Charged Bose Gas*, Phys. Rev. **127**, 1809 (1962).

[Gi5]   M. Girardeau, *Variational method for the quantum statistics of interacting particles*, J. Math. Phys. **3**, 131 (1962).

[GW]   M.D. Girardeau and E.M. Wright, *Bose-Fermi variational Theory for the BEC-Tonks Crossover*, Phys. Rev. Lett. **87**, 210402-1-4 (2001).

[GWT]   M.D. Girardeau, E.M. Wright, and J.M. Triscari, *Ground-state properties of a one-dimensional system of hard-core bosons in a harmonic trap*, Phys. Rev. A **63**, 033601-1-6 (2001).

[Go]   A. Görlitz, *et al.*, *Realization of Bose-Einstein Condensates in Lower Dimension*, Phys. Rev. Lett. **87**, 130402-1-4 (2001).

[GS]   G.M. Graf and J.P. Solovej, *A correlation estimate with applications to quantum systems with Coulomb interactions*, Rev. Math. Phys. **6**, 977–997 (1994).

[G1]   M. Greiner, I. Bloch, O. Mandel, T.W. Hänsch, T. Esslinger, *Exploring Phase Coherence in a 2D Lattice of Bose-Einstein Condensates*, Phys. Rev. Lett. **87**, 160405 (2001).

[G2]   M. Greiner, O. Mandel, T. Esslinger, T.E. Hänsch, I. Bloch, *Quantum phase transition from a superfluid to a Mott insulator in a gas of ultracold atoms*, Nature **415**, 39 (2002).

[G3]   M. Greiner, O. Mandel, T.E. Hänsch, I. Bloch, *Collapse and revival of the matter wave field of a Bose-Einstein condensate*, Nature **419**, 51 (2002).

[Gri]   R.B. Griffiths, *Spontaneous magnetization in idealized ferromagnets*, Phys. Rev. **152**, 240 (1966).

[Gr1]   E.P. Gross, *Structure of a Quantized Vortex in Boson Systems*, Nuovo Cimento **20**, 454–466 (1961).

[Gr2]   E.P. Gross, *Hydrodynamics of a superfluid condensate*, J. Math. Phys. **4**, 195–207 (1963).

[HL]   K. Hepp, E.H. Lieb, *Equilibrium statistical mechanics of matter interacting with the quantized radiation field*, Phys. Rev. A **8**, 2517 (1973).

[HFM]   D.F. Hines, N.E. Frankel, D.J. Mitchell, *Hard disc Bose gas*, Phys. Lett. **68A**, 12–14 (1978).

[Ho]   P.C. Hohenberg, *Existence of Long-range Order in One and Two Dimensions*, Phys. Rev. **158**, 383–386 (1966).

[HoM]   P.C. Hohenberg and P.C. Martin, *Microscopic theory of helium*, Ann. Phys. (NY) **34**, 291 (1965).

[H]   K. Huang, *Bose-Einstein condensation and superfluidity*, in: *Bose-Einstein Condensation*, A. Griffin, D.W. Stroke, S. Stringari, eds., Cambridge University Press, 31–50 (1995).

[HY]   K. Huang, C.N. Yang, Phys. Rev. **105**, 767–775 (1957); T.D. Lee, K. Huang, C.N. Yang, Phys. Rev. **106**, 1135–1145 (1957); K.A. Brueckner, K. Sawada, Phys. Rev. **106**, 1117–1127, 1128–1135 (1957); S.T. Beliaev, Sov. Phys.-JETP **7**, 299–307 (1958); T.T. Wu, Phys. Rev. **115**, 1390 (1959); N. Hugenholtz, D. Pines, Phys. Rev. **116**, 489 (1959); M. Girardeau, R. Arnowitt, Phys. Rev. **113**, 755 (1959); T.D. Lee, C.N. Yang, Phys. Rev. **117**, 12 (1960).

[ISW]   M. Inguscio, S. Stringari, C. Wieman (eds.), *Bose-Einstein Condensation in Atomic Gases*, Italian Physical Society (1999).

[JK]   A.D. Jackson and G.M. Kavoulakis, *Lieb Mode in a Quasi-One-Dimensional Bose-Einstein Condensate of Atoms*, Phys. Rev. Lett. **89**, 070403 (2002).

[JBCGZ]   D. Jaksch, C. Bruder, J.I. Cirac, C.W. Gardiner, P. Zoller, *Cold bosonic atoms in optical lattices*, Phys. Rev. Lett. **81**, 3108–3111 (1998).

[KLS]   T. Kennedy, E.H. Lieb, S. Shastry, *The XY Model has Long-Range Order for all Spins and all Dimensions Greater than One*, Phys. Rev. Lett. **61**, 2582–2584 (1988).

[KD]   W. Ketterle, N.J. van Druten, *Evaporative Cooling of Trapped Atoms*, in: B. Bederson, H. Walther, eds., Advances in Atomic, Molecular and Optical Physics, **37**, 181–236, Academic Press (1996).

[KS]   J. Klauder, B.-S. Skagerstam, *Coherent states, applications in physics and mathematical physics*, World Scientific (1985).

[KNSQ]   E.B. Kolomeisky, T.J. Newman, J.P. Straley, X. Qi, *Low-dimensional Bose liquids: beyond the Gross-Pitaevskii approximation*, Phys. Rev. Lett. **85**, 1146–1149 (2000).

[KT]   M. Kobayashi and M. Tsubota, *Bose-Einstein condensation and superfluidity of a dilute Bose gas in a random potential*, Phys. Rev. B **66**, 174516 (2002).

[KP]   S. Komineas and N. Papanicolaou, *Vortex Rings and Lieb Modes in a Cylindrical Bose-Einstein Condensate*, Phys. Rev. Lett. **89**, 070402 (2002).

[KB]   A.J. Kromminga, M. Bolsterli, *Perturbation Theory of Many-Boson Systems*, Phys. Rev. **128**, 2887 (1962).

[Le]   A. Lenard, *Momentum distribution in the ground state of the one-dimensional system of impenetrable bosons*, J. Math. Phys. **5**, 930–943 (1964).

[Leg]   A.J. Leggett, *Bose-Einstein condensation in the alkali gases: Some fundamental concepts*, Rev. Mod. Phys. **73**, 307 (2001).

[Len]   W. Lenz, *Die Wellenfunktion und Geschwindigkeitsverteilung des entarteten Gases*, Z. Phys. **56**, 778–789 (1929).

[L1]   E.H. Lieb, *Exact Analysis of an Interacting Bose Gas. II. The Excitation Spectrum*, Phys. Rev. **130**, 1616–1624 (1963).

[L2]   E.H. Lieb, *Simplified Approach to the Ground State Energy of an Imperfect Bose Gas*, Phys. Rev. **130**, 2518–2528 (1963). See also Phys. Rev. **133** (1964), A899–A906 (with A.Y. Sakakura) and Phys. Rev. **134** (1964), A312–A315 (with W. Liniger).

[L3]   E.H. Lieb, *The Bose Fluid*, in: W.E. Brittin, ed., Lecture Notes in Theoretical Physics VIIC, Univ. of Colorado Press, pp. 175–224 (1964).

[L4]   E.H. Lieb, *The classical limit of quantum spin systems*, Commun. Math. Phys. **31**, 327–340 (1973).

[L5]   E.H. Lieb, *Thomas-Fermi and related theories of atoms and molecules*, Rev. Mod. Phys. **53**, 603 (1981). Errata **54**, 311 (1982).

[L6]   E.H. Lieb, *Coherent states as a tool for obtaining rigorous bounds*, in: *Coherent States*, D.H. Feng, J. Klauder, M.R. Strayer, eds., p. 267, World Scientific (1994).

[L7]   E.H. Lieb, *The Bose Gas: A Subtle Many-Body Problem*, in: *Proceedings of the XIII International Congress on Mathematical Physics, London*, A. Fokas, et al. eds., pp. 91–111, International Press (2001).

[LL]   E.H. Lieb, W. Liniger, *Exact Analysis of an Interacting Bose Gas. I. The General Solution and the Ground State*, Phys. Rev. **130**, 1605–1616 (1963).

[LLo]   E.H. Lieb, M. Loss, *Analysis*, 2nd ed., Amer. Math. Society, Providence, R.I. (2001).

[LN]   E.H. Lieb, H. Narnhofer, *The Thermodynamic Limit for Jellium*, J. Stat. Phys. **12**, 291–310 (1975). Errata J. Stat. Phys. **14**, 465 (1976).

[LSe]   E.H. Lieb, R. Seiringer, *Proof of Bose-Einstein Condensation for Dilute Trapped Gases*, Phys. Rev. Lett. **88**, 170409-1–4 (2002).

[LSe2]   E.H. Lieb, R. Seiringer, *Derivation of the Gross-Pitaevskii Equation for Rotating Bose Gases*, preprint, arXiv:math-ph/0504042.

[LSSY]   E.H. Lieb, R. Seiringer, J.P. Solovej, and J. Yngvason, *The ground state of the Bose gas*, in: Current Developments in Mathematics, 2001, 131–178, International Press, Cambridge (2002).

[LSSY2]   E.H. Lieb, R. Seiringer, J.P. Solovej, and J. Yngvason, *The ground state of the Bose gas*, in: Proceedings of the conference 'Perspectives in Analysis', KTH Stockholm, 2003, M. Benedicks, P. Jones and S. Smirnov, eds., Springer (in press).

[LSeY1]   E.H. Lieb, R. Seiringer, J. Yngvason, *Bosons in a Trap: A Rigorous Derivation of the Gross-Pitaevskii Energy Functional*, Phys. Rev A **61**, 043602 (2000).

[LSeY2]   E.H. Lieb, R. Seiringer, J. Yngvason, *A Rigorous Derivation of the Gross-Pitaevskii Energy Functional for a Two-dimensional Bose Gas*, Commun. Math. Phys. **224**, 17 (2001).

[LSeY3]   E.H. Lieb, R. Seiringer, J. Yngvason, *The Ground State Energy and Density of Interacting Bosons in a Trap*, in: *Quantum Theory and Symmetries*, Goslar, 1999, H.-D. Doebner, V.K. Dobrev, J.-D. Hennig and W. Luecke, eds., pp. 101–110, World Scientific (2000).

[LSeY4]   E.H. Lieb, R. Seiringer, J. Yngvason, *Two-Dimensional Gross-Pita-evskii Theory*, in: Progress in Nonlinear Science, Proceedings of the International Conference Dedicated to the 100th Anniversary of A.A. Andronov, Volume II, A.G. Litvak, ed., pp. 582–590, Nizhny Novgorod, Institute of Applied Physics, University of Nizhny Novgorod (2002).

[LSeY5]   E.H. Lieb, R. Seiringer, J. Yngvason, *Superfluidity in Dilute Trapped Bose Gases*, Phys. Rev. B **66**, 134529 (2002).

[LSeY6]   E.H. Lieb, R. Seiringer, J. Yngvason, *One-Dimensional Behavior of Dilute, Trapped Bose Gases*, Commun. Math. Phys. **244**, 347–393 (2004). See also: *One-Dimensional Bosons in Three-Dimensional Traps*, Phys. Rev. Lett. **91**, 150401-1–4 (2003).

[LSeY7]   E.H. Lieb, R. Seiringer, and J. Yngvason, *Poincaré Inequalities in Punctured Domains*, Ann. Math. **158**, 1067–1080 (2003).

[LSeY8]   E.H. Lieb, R. Seiringer, and J. Yngvason, *Justification of c-Number Substitutions in Bosonic Hamiltonians*, Phys. Rev. Lett. **94**, 080401 (2005).

[LSo]   E.H. Lieb, J.P. Solovej, *Ground State Energy of the One-Component Charged Bose Gas*, Commun. Math. Phys. **217**, 127–163 (2001). Errata **225**, 219–221 (2002).

[LSo2]   E.H. Lieb, J.P. Solovej, *Ground State Energy of the Two-Component Charged Bose Gas*, Commun. Math. Phys. **252**, 485–534 (2004).

[LT]   E.H. Lieb, L.E. Thomas, *Exact ground state energy of the strong-coupling polaron*, Commun. Math. Phys. **183**, 511 (1997). Errata *ibid* **188**, 499 (1997).

[LY1]   E.H. Lieb, J. Yngvason, *Ground State Energy of the low density Bose Gas*, Phys. Rev. Lett. **80**, 2504–2507 (1998).

[LY2]   E.H. Lieb, J. Yngvason, *The Ground State Energy of a Dilute Two-dimensional Bose Gas*, J. Stat. Phys. **103**, 509 (2001).

[LY3]   E.H. Lieb, J. Yngvason, *The Ground State Energy of a Dilute Bose Gas*, in: *Differential Equations and Mathematical Physics, University of Alabama, Birmingham, 1999*, R. Weikard and G. Weinstein, eds., pp. 271–282, Amer. Math. Soc./Internat. Press (2000).

[Lu]   M. Luban, *Statistical Mechanics of a Nonideal Boson Gas: Pair Hamiltonian Model*, Phys. Rev. **128**, 965 (1962).

[Ma]   P.C. Martin, *A microscopic approach to superfluidity and superconductivity*, J. Math. Phys. **4**, 208 (1963).

[MM]   T. Matsubara, H. Matsuda, *A lattice model of liquid helium*, Progr. Theor. Phys. **16**, 569–582 (1956).

[MW]   N.D. Mermin, H. Wagner, *Absence of Ferromagnetism or Antiferromagnetism in One- or Two-Dimensional Isotropic Heisenberg Models*, Phys. Rev. Lett. **17**, 1133–1136 (1966).

[MSKE]   H. Moritz, T. Stöferle, M. Köhl and T. Esslinger, *Exciting Collective Oscillations in a Trapped 1D Gas*, Phys. Rev. Lett. **91**, 250402 (2003).

[MA]   O. Morsch, E. Arimondo, *Ultracold atoms and Bose-Einstein condensates in optical lattices*, Lecture Notes in Physics Vol. 602, Springer (2002).

[M]   W.J. Mullin, *Bose-Einstein Condensation in a Harmonic Potential*, J. Low Temp. Phys. **106**, 615–642 (1997).

[NS]   Z. Nazario, D.I. Santiago, *Quantum states of matter of simple bosonic systems: BECs, superfluids and quantum solids*, Phys. Lett. A **328**, 207–211 (2004).

[Ol]   M. Olshanii, *Atomic Scattering in the Presence of an External Confinement and a Gas of Impenetrable Bosons*, Phys. Rev. Lett. **81**, 938–941 (1998).

[OS]   K. Osterwalder, R. Schrader, *Axioms for Euclidean Green's Functions*, Commun. Math. Phys. **31**, 83–112 (1973); Commun. Math. Phys. **42**, 281–305 (1975).

[O]   A.A. Ovchinnikov, *On the description of a two-dimensional Bose gas at low densities*, J. Phys. Condens. Matter **5**, 8665–8676 (1993). See also JETP Letters **57**, 477 (1993); Mod. Phys. Lett. **7**, 1029 (1993).

[PO]   O. Penrose, L. Onsager, *Bose-Einstein Condensation and Liquid Helium*, Phys. Rev. **104**, 576–84 (1956).

[PS]   C. Pethick, H. Smith, *Bose-Einstein Condensation of Dilute Gases*, Cambridge University Press (2001).

[PHS]   D.S. Petrov, M. Holzmann and G.V. Shlyapnikov, *Bose-Einstein Condensates in Quasi-2D trapped Gases*, Phys. Rev. Lett. **84**, 2551 (2000).

[PSW]   D.S. Petrov, G.V. Shlyapnikov, and J.T.M. Walraven, *Regimes of Quantum Degeneracy in Trapped 1D Gases*, Phys. Rev. Lett. **85**, 3745–3749 (2000).

[Pi]   L.P. Pitaevskii, *Vortex lines in an imperfect Bose gas*, Sov. Phys. JETP. **13**, 451–454 (1961).

[PiSt]   L. Pitaevskii, S. Stringari, *Uncertainty Principle, Quantum Fluctuations, and Broken Symmetries*, J. Low Temp. Phys. **85**, 377 (1991).

[PiSt2]   L. Pitaevskii and S. Stringari, *Bose-Einstein Condensation*, Oxford Science Publications, Oxford (2003).

[Po]   V.N. Popov, *On the theory of the superfluidity of two- and one-dimensional Bose systems*, Theor. and Math. Phys. **11**, 565–573 (1977).

[PrSv]   N.V. Prokof'ev and B.V. Svistunov, *Two definitions of superfluid density*, Phys. Rev. B **61**, 11282 (2000).

[R]   A.M. Rey, K. Burnett, R. Roth, M. Edwards, C.J. Williams, C.W. Clark, *Bogoliubov approach to superfluidity of atoms in an optical lattice*, J. Phys. B **36**, 825–841 (2003).

[Ro]   G. Roepstorff, *Bounds for a Bose condensate in dimensions $\nu \geq 3$*, J. Stat. Phys. **18**, 191 (1978).

[Ru]   D. Ruelle, *Statistical Mechanics. Rigorous Results*, World Scientific (1999).

[Sa]   S. Sachdev, *Quantum Phase Transitions*, Cambridge University Press (1999).

[S]   M. Schick, *Two-Dimensional System of Hard Core Bosons*, Phys. Rev. A **3**, 1067–1073 (1971).

[SY]   K. Schnee, J. Yngvason, *Bosons in disc-shaped traps: from 3D to 2D*, preprint.

[Sc]   F. Schreck, *et al.*, *Quasipure Bose-Einstein Condensate Immersed in a Fermi Sea*, Phys. Rev. Lett. **87**, 080403 (2001).

[Se1]   R. Seiringer, Diplom thesis, University of Vienna (1999).

[Se2]   R. Seiringer, *Bosons in a Trap: Asymptotic Exactness of the Gross-Pitaevskii Ground State Energy Formula*, in: *Partial Differential Equations and Spectral Theory*, PDE2000 Conference in Clausthal, Germany, M. Demuth and B.-W. Schulze, eds., 307–314, Birkhäuser (2001).

[Se3]   R. Seiringer, *Gross-Pitaevskii Theory of the Rotating Bose Gas*, Commun. Math. Phys. **229**, 491–509 (2002).

[Se4]   R. Seiringer, *Ground state asymptotics of a dilute, rotating gas*, J. Phys. A: Math. Gen. **36**, 9755–9778 (2003).

[Sh]   S.I. Shevchenko, *On the theory of a Bose gas in a nonuniform field*, Sov. J. Low Temp. Phys. **18**, 223–230 (1992).

[Si1]   B. Simon, *Trace ideals and their application*, Cambridge University Press (1979).

[Si2]   B. Simon, *The classical limit of quantum partition functions*, Commun. Math. Phys. **71**, 247 (1980).

[So]   J.P. Solovej, *Upper Bounds to the Ground State Energies of the One- and Two-Component Charged Bose gases*, preprint, arXiv:math-ph/0406014.

[SR]   L. Spruch, L. Rosenberg, *Upper bounds on scattering lengths for static potentials*, Phys. Rev. **116**, 1034 (1959).

[Su]   A. Sütő, *Bose-Einstein condensation and symmetry breaking*, Phys. Rev. A **71**, 023602 (2005).

[Su2]   A. Sütő, *Equivalence of Bose-Einstein Condensation and Symmetry Breaking*, Phys. Rev. Lett. **94**, 080402 (2005).

[T]   G. Temple, *The theory of Rayleigh's Principle as Applied to Continuous Systems*, Proc. Roy. Soc. London A **119**, 276–293 (1928).

[Th]   W. Thirring, *A lower bound with the best possible constants for Coulomb Hamiltonians*, Commun. Math. Phys. **79**, 1 (1987).

[TT]   D.R. Tilley and J. Tilley, *Superfluidity and Superconductivity*, third edition, Adam Hilger, Bristol and New York (1990).

[W]   G. Wentzel, *Thermodynamically Equivalent Hamiltonian for Some Many-Body Problems*, Phys. Rev. **120**, 1572 (1960).

[Yu]   V.I. Yukalov, *Principal problems in Bose-Einstein condensation of dilute gases*, Laser Phys. Lett. **1**, 435–461 (2004).

[ZB]   V.A. Zagrebnov, J.B. Bru *The Bogoliubov model of weakly interacting Bose gas*, Phys. Reports **350**, 291–434 (2001).

[ZFG]   W.-M. Zhang, D.H. Feng, R. Gilmore, *Coherent states: Theory and some applications*, Rev. Mod. Phys. **62**, 867 (1990).

[Z1]   K. Ziegler, *Phase Transition of a Bose Gas in an Optical Lattice*, Laser Physics **13**, 587–593 (2003).

[Z2]   K. Ziegler, *Two-component Bose gas in an optical lattice at single-particle filling*, Phys. Rev. A **68**, 053602 (2003).

[Zw]   W. Zwerger, *Mott-Hubbard transition of cold atoms in optical lattices*, Journal of Optics B **5**, 9–16 (2003).

Index